Lecture Notes in Artificial Intelligence　10312

Subseries of Lecture Notes in Computer Science

LNAI Series Editors

Randy Goebel
  University of Alberta, Edmonton, Canada
Yuzuru Tanaka
  Hokkaido University, Sapporo, Japan
Wolfgang Wahlster
  DFKI and Saarland University, Saarbrücken, Germany

LNAI Founding Series Editor

Joerg Siekmann
  DFKI and Saarland University, Saarbrücken, Germany

More information about this series at http://www.springer.com/series/1244

Annalisa Appice · Michelangelo Ceci
Corrado Loglisci · Elio Masciari
Zbigniew W. Raś (Eds.)

New Frontiers
in Mining
Complex Patterns

5th International Workshop, NFMCP 2016
Held in Conjunction with ECML-PKDD 2016
Riva del Garda, Italy, September 19, 2016
Revised Selected Papers

 Springer

Editors
Annalisa Appice
Università degli Studi di Bari Aldo Moro
Bari
Italy

Michelangelo Ceci
Università degli Studi di Bari Aldo Moro
Bari
Italy

Corrado Loglisci
Università degli Studi di Bari Aldo Moro
Bari
Italy

Elio Masciari
ICAR-CNR
Rende
Italy

Zbigniew W. Raś
University of North Carolina
Charlotte, NC
USA

ISSN 0302-9743   ISSN 1611-3349   (electronic)
Lecture Notes in Artificial Intelligence
ISBN 978-3-319-61460-1   ISBN 978-3-319-61461-8   (eBook)
DOI 10.1007/978-3-319-61461-8

Library of Congress Control Number: 2017944352

LNCS Sublibrary: SL7 – Artificial Intelligence

Printed on acid-free paper

This Springer imprint is published by Springer Nature
The registered company is Springer International Publishing AG
The registered company address is: Gewerbestrasse 11, 6330 Cham, Switzerland

Preface

Modern automatic systems are able to collect huge volumes of data, often with a complex structure (e.g., multi-table data, XML data, Web data, time series and sequences, graphs, and trees). This fact poses new challenges for current information systems with respect to storing, managing, and mining these big sets of complex data.

The 5th International Workshop on New Frontiers in Mining Complex Patterns (NFMCP 2016) was held in Riva del Garda in conjunction with the European Conference on Machine Learning and Principles and Practice of Knowledge Discovery in Databases (ECML-PKDD 2016) on September 19, 2016. The purpose of this workshop was to bring together researchers and practitioners in data mining who are interested in the advances and latest developments in the area of extracting patterns from big and complex data sources. The workshop was aimed at integrating recent results from existing fields such as data mining, statistics, machine learning, and relational databases to discuss and introduce new algorithmic foundations and representation formalisms in complex pattern discovery.

This book features a collection of revised and significantly extended versions of papers accepted for presentation at the workshop. These papers went through a rigorous review process to ensure compliance with Springer's high-quality publication standards. The individual contributions of this book illustrate advanced data-mining techniques that preserve the informative richness of complex data and allow for efficient and effective identification of complex information units present in such data.

The book is composed of five parts and a total of 16 chapters.

Part I analyzes Feature Selection and Induction in the presence of complex data. It consists of two chapters. Chapter 1 introduces an unsupervised algorithm for feature construction based on tree ensembles. It defines an informative data representation that is able to handle complex data structures, combining information from multiple sources. Chapter 2 presents a graph-based algorithm for feature selection. It ranks features by identifying the most important ones into an arbitrary set of cues.

Part II focuses on Classification and Prediction by illustrating some complex predictive problems. It consists of five chapters. Chapter 3 tackles the problem of pruning rule classifiers, while retaining their descriptive properties. It uses confirmation measures as representatives of interestingness measures designed to select rules with desirable descriptive properties. Chapter 4 studies the problem of automatically recognizing speed changes from audio data recorded in controlled conditions. The classification of the audio data is performed using random forests, deep learning architectures and support vector machines. Chapter 5 describes a classification task that aims at determining whether two voices are spoken by the same person or not. It illustrates an algorithm that performs the classification by evaluating the dissimilarity between a speech sample and a set of known models. Chapter 6 investigates the problem of interpreting rules induced from imbalanced data. It proposes three different strategies that combine Bayesian confirmation measures, in order to select rules having good descriptive characteristics. Chapter 7 addresses the problem of modeling trust network evolution through social communications among users in a social media site. It introduces a link prediction algorithm based on mediating-objects and analyzes the effect of time-decay in creating trust-links.

Part III analyzes issues posed by Clustering in the presence of complex data. It consists of four chapters. Chapter 8 investigates the adoption of cluster analysis to predict the primary medical procedure for a patient. The processed patients are clustered according to their set of diagnoses. This cluster knowledge is then used to identify other existing patients that are considered similar to the new patient. Chapter 9 describes a clustering algorithm allowing us to group features that are likely to take extreme values simultaneously. It exploits the graphical structure stemming from the definition of the clusters. Chapter 10 presents a latent-factor-based approach whose goal is to profile users according to their behavior. It considers the actions as set of features instead of single atomic elements. Chapter 11 proposes a multiview clustering methodology that determines clusters of patients with similar symptoms and detects patterns of medication changes that lead to the improvement or decline of patients' quality of life.

Part IV presents algorithms Pattern Discovery. It consists of three chapters. Chapter 12 introduces an approach to extract recurrent deviations from historical logging data and generate anomalous patterns representing high-level deviations. It applies a frequent subgraph mining technique together with an ad hoc conformance-checking technique. Chapter 13 investigates the task of detecting weather changes, which are periodically repeated over time and space. It introduces a spatiotemporal pattern to represent a periodic change and describes a computational solution to discover this kind of pattern. Chapter 14 investigates the problem of user authentication based on keystroke timing pattern. It proposes a simple, robust, and nonparameterized nearest-neighbor regression-based feature-ranking algorithm for anomaly detection.

Finally, Part V gives a general overview of Applications in sensor network and game scenarios. It contains two chapters. Chapter 15 provides a formalization of a graph-based approach that extends a directed weighted graph using a sequential state transformation function. It interprets the graph to model state transition matrices and describes an algorithm for deriving these interpretations in large-scale real-world sensor networks. Chapter 16 checks whether, and to what extent, advanced process mining techniques can support efficient and effective knowledge discovery in chess playing. It also provides interesting insight into the game rules and strategies, and/or may support effective game playing in future matches.

We would like to thank all the authors who submitted papers for publication in this book and all the workshop participants and speakers. We are also grateful to the members of the Program Committee and additional reviewers for their excellent work in reviewing submitted and revised contributions with expertise and patience. We would like to thank Jaakko Hollmen for his invited talk on "On Model, Patterns, and Prediction." A special thanks is due to both the ECML PKDD Workshop Chairs and to the ECML PKDD organizers who made the event possible. We would like to acknowledge the support of the European Commission through the projects

MAESTRA–Learning from Massive, Incompletely Annotated, and Structured Data (Grant number ICT-2013-612944) and TOREADOR–Trustworthy Model-Aware Analytics Data Platform (Grant number H2020-688797). Last but not the least, we thank Alfred Hofmann of Springer for his continuous support.

March 2017

Annalisa Appice
Michelangelo Ceci
Corrado Loglisci
Elio Masciari
Zbigniew Ras

Organization

Program Chairs

Annalisa Appice   University of Bari Aldo Moro, Bari, Italy
Michelangelo Ceci   University of Bari Aldo Moro, Bari, Italy
Corrado Loglisci   University of Bari Aldo Moro, Bari, Italy
Elio Masciari   ICAR-CNR, Rende, Italy
Zbigniew Ras   University of North Carolina, Charlotte, USA and Warsaw
         University of Technology, Poland

Program Committee

Nicola Barbieri   Yahoo Research Barcelona, Spain
Elena Bellodi   University of Ferrara, Italy
Petr Berka   University of Economics, Prague, Czech Republic
Jorge Bernardino   Polytechnic Institute of Coimbra, Portugal
Claudia Diamantini   Polytechnic University of Marche, Italy
Bettina Fazzinga   CNR-ICAR, Italy
Stefano Ferilli   University of Bari, Italy
Hongyu Guo   National Research Council Canada, Canada
Dino Ienco   IRSTEA, France
Dragi Kocev   Jozef Stefan Institute, Slovenia
Ruggero G. Pensa   University of Turin, Italy
Nhathai Phan   University of Oregon, USA
Nico Piatkowski   Technische Universität Dortmund, Germany
Gianvito Pio   University of Bari, Italy
Domenico Redavid   Artificial Brain, Italy
Rita P. Ribeiro   University of Porto, Portugal
Jerzy Stefanowski   Poznan University of Technology, Poland
Herna Viktor   University of Ottawa, Canada
Alicja Wieczorkowska   Polish-Japanese Institute of Information Technology,
         Poland
Wlodek Zadrozny   University of North Carolina, Charlotte USA

Additional Reviewers

Angelastro, Sergio   Mileski, Vanja
Branco, Paula   Mircoli, Alex
Ferilli, Stefano   Pazienza, Andrea
Genga, Laura

On Models, Patterns, and Prediction
(Invited Talk)

Jaakko Hollmen

Department of Computer Science, Aalto University in Espoo, Espoo, Finland

Abstract. Pattern discovery has been the center of attention of data mining research for a long time, with patterns languages varying from simple to complex, according to the needs of the applications and the format of data. In this talk, I will take a view on pattern mining that combines elements from neighboring areas. More specifically, I will describe our previous research work in the intersection of the three areas: probabilistic modeling, pattern mining and predictive modeling. Clustering in the context of pattern mining will be explored, as well as linguistic summarization of patterns. Also, multiresolution pattern mining as well as semantic pattern discovery and pattern visualization will be visited. Time allowing, I will speak about patterns of missing data and it simplications on predictive modeling.

Contents

Pattern Discovery

Applications

Feature Selection and Induction

Feature Induction and Network Mining with Clustering Tree Ensembles

Konstantinos Pliakos$^{(\boxtimes)}$ and Celine Vens

Department of Public Health and Primary Care, KU Leuven, Campus KULAK,
Etienne Sabbelaan 53, 8500 Kortrijk, Belgium
{konstantinos.pliakos,celine.vens}@kuleuven.be

Abstract. The volume of data generated and collected using modern technologies grows exponentially. This vast amount of data often follows a complex structure, and the problem of efficiently mining and analyzing such data is crucial for the performance of various machine learning tasks. Here, a novel data mining framework for unsupervised learning tasks is proposed based on decision tree learning and ensembles of trees. The proposed approach introduces an informative feature representation and is able to handle data diversity and complexity. Moreover, a new scheme is proposed based on the aforementioned approach for mining interaction data. These data are often modeled as homogeneous or heterogeneous networks and they are present in various fields, such as social media, recommender systems, and bioinformatics. The learning process is performed in an unsupervised manner, following also the inductive setup. The experimental evaluation confirms the effectiveness of the proposed approach.

Keywords: Tree-ensembles · Extremely randomized trees · Tree-embedding · Network mining

1 Introduction

Nowadays, a great advance in data acquisition and feature construction methods is witnessed. Due to modern technological advances, huge amounts of data are generated in terms of both cardinality (i.e., the number of samples) and dimensionality (i.e., the number of features that describe each sample). These data often follow more complex structures, combining information from multiple sources. One example that is often encountered is interaction data. Instead of one set of objects described by a set of features, interaction data is characterized by two sets of objects, each described by its own set of features. Interaction data is omni-present: in social network analysis, recommender systems, ecology (habitat modeling), bioinformatics (gene expression analysis, drug response analysis, predicting drug-target reactions), technology-enhanced education, etc. Furthermore, as the volume of data grows, problems such as the existing noise in the data or the missing values in some datasets remain. To this end, methods that

© Springer International Publishing AG 2017
A. Appice et al. (Eds.): NFMCP 2016, LNAI 10312, pp. 3–18, 2017.
DOI: 10.1007/978-3-319-61461-8_1

can handle the aforementioned issues and succeed in mining complex patterns in big datasets are indisputably needed.

During the last years, an interest was witnessed in leveraging the mining of complex patterns by mapping the data to different feature spaces. This way, the performance of machine learning algorithms was improved. Most of the developed methods were based on kernel learning [1,2], mainly due to the very good performance of Support Vector Machines (SVMs) [3]. However, these methods are often characterized by high computational costs and limited flexibility as one should compute and handle the whole Gram matrix. Many of these kernel-based methods have also been developed in a transductive setup where test instances are available during the training phase [1].

There are several studies where new features are constructed inductively using clustering techniques or decision tree learning. Most of the recently developed feature construction methods were developed for supervised learning tasks. In [4], a feature induction method based on random forests [5] was proposed. It was based on a metric transformation that mapped the identity of the tests performed in each node of a decision tree to a feature indicator. Feature vectors were generated by concatenating all the features corresponding to each tree in the forest and they were further encoded using hashing. A similar transformation of the data, using a set of random clustering forests was proposed in [6,7] for visual codebook construction. In particular, the features were generated by randomized trees. The data encoding was based only on the indices of the leaves where a data sample ends up. The approach leads to a high dimensional, sparse binary coding. In [8], a label-specific feature scheme for multi-label classification was proposed. For each label, a distinct feature set was constructed by clustering the label's positive and negative instances (separately), and then calculating the distances of each instance to the obtained cluster centroids. This way, the predictive performance of a classifier trained for that specific label was increased.

Here, we focus on developing a feature representation using tree ensembles. The main goal is to leverage unsupervised machine learning tasks, such as clustering or information retrieval. Decision tree induction algorithms [9,10] are among the most popular data mining algorithms. They have been applied extensively in many fields such as systems biology [11] or social media analysis [12]. The interpretability of the models they produce is among the main advantages of these methods, making them transparent and understandable to human experts, also leveraging knowledge discovery. Other advantages include their scalability from a computational point of view and their fair predictive accuracy. Combining them with ensemble methods [5,13] improves their predictive performance and provides state-of-the-art results.

Motivated by [4], here we propose an unsupervised framework for feature construction based on tree ensembles and specifically Extremely Randomized Trees [14], hereafter denoted as *ERT*. In particular, the nodes of each decision tree of the ensemble are treated as clusters, containing all the samples that fall into that tree node. Next, binary feature vectors are generated, where each component represents the presence or absence of a sample in a cluster (node).

The new features are generated in an inductive manner (i.e., the test samples are not needed during learning). Different from [4], the learning procedure is performed in an unsupervised manner. In addition, the employment of dimensionality reduction techniques [15,16] is studied and the efficiency in detecting an underlying manifold over complex data is tested.

Furthermore, the proposed data representation approach is extended towards interaction data. Relations between entities that interact with each other such as user-item relations in recommender systems or drug-patient interactions in medicine are often represented by networks (here, equally referred to as graphs). Generally, there are two types of networks, homogeneous that model samples of the same type (e.g., protein-protein network) and bi-partite modeling samples of different type (e.g., drug-protein network). Despite the continuous rising in the amount of available data, usually we have only a very partial knowledge of these networks [17]. Both supervised and unsupervised machine learning methods have been used to complete a partially known network or to reveal unprecedented knowledge by extracting existing patterns from it [18,19]. There are mainly two methodologies to apply a learning technique in the aforementioned framework, the local approach [20] and the global one [21]. Following the local approach one should first decompose the data into separate (traditional) feature vector representations, solve each representation 's learning task independently, and combine the results. In the global approach, the learning technique is adapted so that it can handle the structured representation directly. In [17], the global approach was based on building a global representation of the network and then treat the interaction prediction problem as a binary classification task. Here, a method is proposed that combines these two approaches in a unified framework. More precisely, the aforementioned feature induction approach based on ERT is applied on each set of the two interacting entities separately (local part), producing two new high-dimensional sparse representations. Next, after transferring the two sets to lower dimensional spaces we combine the two separate low-dimensional feature representations, building this way a global representation of the network. To this end, it can be concluded that the proposed approach yields a new global network representation that is more informative and computationally more efficient. The experimental results demonstrate the effectiveness of the proposed approach.

The outline of the paper is as follows. In Sect. 2, the proposed approach is described in detail. The experimental evaluation is presented in Sect. 3. Conclusions are drawn and topics of future research are discussed in Sect. 4.

2 Method

2.1 Learning Using Extremely Randomized Trees

Decision trees are typically constructed with a top-down induction method. Starting from the root node that is associated with the complete training set, the nodes are recursively split by applying a test to one of the features. In order to find the best split, all features and their corresponding split points are considered and a split quality criterion is evaluated. In supervised learning tasks, this criterion is often information gain (classification), or variance reduction (regression). When the data contained in a node is pure w.r.t. the target, or when some other stopping criterion holds, the node becomes a leaf node and a prediction is assigned to it. This prediction is the majority class assigned to the training instances in the leaf for classification, or the average of their target values for regression. The prediction for test instances is obtained by sorting them through the tree into a leaf node. In this work, the decision tree learners employed are set in the Predictive Clustering Tree (PCT) [10] framework, adopting the hierarchical clustering view of decision trees. PCTs are constructed by maximally reducing intra-cluster variance at each split. By computing the variance over the feature set, rather than the target, PCTs can be applied to (unsupervised) clustering tasks.

Since decision trees often have a large variance, their predictive performance can be improved by having several trees returning an aggregated prediction. Such a collection of decision trees is called an ensemble, and several instances of ensembles exist. In this work, we consider the ensemble method of Extremely Randomized Trees (ERT) [14,22]. The ERT algorithm builds an ensemble of unpruned decision trees following the traditional top-down procedure. In an ERT ensemble, each tree is constructed by considering only a random set of split candidates at each node. More precisely, a random subset of features is picked, and for each feature, a random split point is picked. From these candidates, the candidate yielding the best value for the split criterion is chosen. The growing of each tree is stopped when the tree is fully grown (i.e., one sample in each leaf) or a criterion has been reached (e.g., maximum depth, minimum number of samples to split, etc.). The rationale behind the ERT algorithm is that the explicit randomization of the splitting threshold and attribute in combination with ensemble averaging reduces bias-variance more strongly than the randomization performed by other methods. ERT was shown to have a better predictive performance than the more popular Random Forests [14] and it is also computationally less expensive due to the simplicity of the node splitting procedure.

2.2 Feature Construction with Extremely Randomized Trees

A new feature set is generated by applying ERT on the initial feature set, as follows. The nodes of each tree in the ERT setting, $\mathbf{C} = \{c_1, c_2, \cdots, c_{|C|}\}$ are treated as clusters containing all the samples that fall into them traversing the tree. Clearly, there is no point into including the root nodes in the procedure. Let $\mathbf{X} \in \Re^{|S| \times |M|}$ be the dataset and $\mathbf{F} \in \Re^{|S| \times |C|}$ the induced feature set, where $|S|$, $|M|$ and $|C|$ correspond to the number of samples, the number of original features, and the number of induced features of the dataset, respectively. Next, the clusters $c_j \in \mathbf{C}$ are treated as features of the feature set $\mathbf{F}$. Each $f_{ij} \in \mathbf{F}$ equals to 1 if the sample $i \in \mathbf{S}$ is contained in the cluster (node) c_j and 0

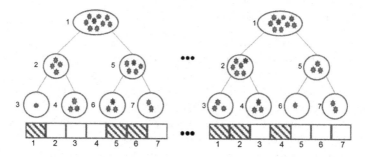

Fig. 1. Illustration of the proposed approach. The example associated with the induced feature vector is depicted as red. (Color figure online)

otherwise. The proposed approach is coined as ERCP (Extremely Randomized Clustering tree-Paths). In Fig. 1, the feature induction approach is shown.

The proposed feature representation is rationally more informative than the original one. Due to the feature selection mechanism of the ERT, features that contain redundant information are not included in the procedure (i.e., no split occurs on these features). The induced features are generated by computing clusterings over the whole dataset and therefore information from the whole instance space is exploited. Samples that are outliers in the dataset can be discriminated easily, as splitting an outlier from the rest of the dataset rationally leads to large variance reduction. In addition, regions of the instance space with high variance will lead to longer paths in the trees, thereby making the procedure adaptive towards the difficulty of the instances considered. Moreover, one can control the growing of the trees by setting specific stopping criteria.

At this point, it has to be noted that a similar encoding could be produced by any hierarchical clustering method. However, the employment of ERT is beneficial. First, ERT is a tree ensemble method, and therefore it is robust to small perturbations in the data. It is also robust to non-informative or noisy features due to the implicit feature selection mechanism. This way, the generated feature representation is considered more noise invariant. Moreover, another advantage is that the tree ensembles can generally treat both numerical and non-numerical values, making the method more easily applied and robust. In addition to that, in contrast to many other methods, it offers a natural way to deal with missing values by distributing instances with a missing split value over all branches or by selecting at random one branch to follow. Other advantages of the proposed approach is that it is parameter-free and it is performed in an inductive manner. After the training, the model can handle any new data without any need of the training set. Furthermore, it is expected that a greater number of examples will lead to bigger trees in the forest. The proposed representation will be therefore larger but also very sparse. This way, the application of our approach to modern online systems as well as systems that handle large scale data is feasible.

2.3   Mining Interaction Data

As mentioned before, the relations between two entities that interact with each other are often represented as a network (here, equally referred to as a graph). Let G define a network with two finite sets of nodes $N_r = \{n_{r1}, \cdots, n_{r|N_r|}\}$ and $N_q = \{n_{q1}, \cdots, n_{q|N_q|}\}$. Each node of the network is described by a feature representation. The network corresponds to a bipartite graph over the two sets of nodes N_r and N_q. The interactions between N_r and N_q are modeled as edges connecting the nodes and are represented in the adjacency matrix $\mathbf{Y} \in \Re^{|N_r| \times |N_q|}$. Every item $y(i, j) \in \mathbf{Y}$ is equal to 1 if an interaction between items n_{ri} and n_{qj} exists and 0 otherwise. Homogeneous graphs defined on only one type of nodes can be obtained as a particular case of the aforementioned general framework by considering two identical sets of nodes (i.e., $N_r = N_q$).

In the proposed approach the bipartite graph is first decomposed into two separate sets of nodes. For example in a drug-protein interaction network one has a set of nodes corresponding to drugs and one corresponding to proteins. Each set of nodes N_r or N_q is represented by a feature set $\mathbf{X_r} \in \Re^{|N_r| \times |M_r|}$ or $\mathbf{X_q} \in \Re^{|N_q| \times |M_q|}$, respectively. Next, two feature sets $\mathbf{F_r} \in \Re^{|N_r| \times |C_r|}$ and $\mathbf{F_q} \in \Re^{|N_q| \times |C_q|}$ are induced by applying ERCP on $\mathbf{X_r}$ and $\mathbf{X_q}$ respectively, as described in Sect. 2.2. The new high dimensional feature representation of the nodes is then transferred to a lower dimensional space d $(d_r \ll |C_r|, d_q \ll |C_q|)$. This transformation could be performed by embedding the data into a linear or non-linear subspace of lower dimensionality. Although many techniques exist, here the most popular Principal Components Analysis (PCA) was employed. By applying PCA the inductive setup of the method is maintained. Next, a global data representation is built as the cartesian product of the two feature spaces. More precisely, a feature vector is generated for each pair of nodes as the concatenation of the feature vectors corresponding to the nodes of each pair. To this end, a global representation $\mathbf{F'}$ is yielded, where $\mathbf{F'} \in \Re^{|||N_r|*|N_q|||\times|||d_r|+|d_q|||}$. In Fig. 2, the proposed model for mining interaction data is displayed.

3   Experimental Evaluation

3.1   Datasets

The evaluation procedure of the proposed approach starts by employing some well-known datasets from UCI repository [23] in order to reveal the global potential of our approach. The evaluation continues by employing more complex datasets and specifically two datasets that correspond to homogeneous biological networks. Next, the evaluation of the interaction data mining approach (Sect. 2.3) follows. Including several datasets from various fields contributes in avoiding any biased conclusions and revealing the robustness of our method. The labels contained in these datasets were used only for evaluation purposes and were not included in any part of the learning process. In Table 1, further information about the used datasets is provided. A pre-processing step was also introduced as in [4]. In particular, the data have been whitened by normalizing

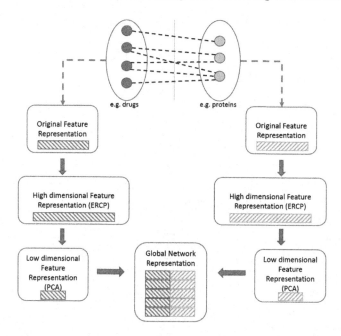

Fig. 2. A description of the proposed interaction data representation model.

all features to have zero mean and unit standard deviation. Non-binary classification tasks were transformed into binary ones by considering the majority class versus all the others or by grouping the classes in two sets of balanced size. Despite the fact that tree-ensembles do not require any pre-processing of the data, in order to compare the proposed feature representation to the original one the missing values were replaced by the features 's average and the nominal features in some datasets were transformed into a set of binary ones using one-hot encoding. This way, algorithms that can not handle missing values (e.g., k-NN, k-means) can be applied on both data representations (original features, induced features) for comparison purposes.

In order to prove the efficiency of the proposed feature representation approach on more complex data structures, 5 interaction prediction datasets [17] were also introduced. They are interaction datasets representing homogeneous and heterogeneous biological networks. In particular:

– **Metabolic network (MN)** [24]. This homogeneous network consists of 668 S. cerivisiae enzymes and the connections represent the catalysation of successive reactions between two enzymes. The enzymes are originally represented by 325 features. They are a set of expression data, phylogenetic profiles and localization data.
– **Protein-protein interaction network (PPI)** [25]. This homogeneous network contains interactions between 984 S. cerivisiae proteins. The input features are also a set of expression data, phylogenetic profiles and localization data.

Table 1. The datasets used in the evaluation procedure.

Dataset	Nb of instances	Nb of features
Pima Indians diabetes	768	8
Ecoli	336	7
Glass identification	163	9
Haberman's survival	306	3
Ionosphere	351	34
Iris	150	4
Libras movement	192	90
Robot execution failures (Lp5)	164	90
Mammographic mass	961	14
Sonar	208	60
Spectf heart	267	44
Statlog (vehicle)	846	18
Breast cancer (orig.)	699	9
Breast cancer (diag.)	569	30
Wine	178	13
Breast cancer (prog.)	198	32

– **E. coli regulatory network (ERN)** [26]. This heterogeneous network consists of 179256 pairs of 154 transcription factors (TF) and 1164 genes of E. coli ($154 \times 1164 = 179256$). The feature vectors that represent the two sets consist of 445 expression values.
– **S. cerevisiae regulatory network (SRN)** [27]. This heterogeneous network is composed by interactions between TFs and their target S. cerevisiae genes. It is composed of 205773 pairs of 1821 genes and 113 TFs. The input features are 1685 expression values. For genes, motifs features were concatenated to the expression values yielding feature vectors of 9884 values.
– **Drug–protein interaction network (DPI)** [28]. In this heterogeneous network a drug is connected with a protein when the drug targets the protein. This network contains interactions between 683 proteins and 1779 drugs, yielding a set of 1215057 pairs. The input feature vectors represent the presence or absence of 660 chemical substructures for each drug, and the presence or absence of 876 PFAM domains for each protein.

3.2 Experimental Results

Although we target unsupervised learning tasks, datasets with known class labels were used in order to better evaluate the proposed feature construction technique, denoted as Extremely Randomized Clustering tree Paths (ERCP). In particular, the class labels were used only as ground truth during evaluation and were disregarded during the learning phase. The performance of a k-NN algorithm applied on the induced features generated by ERCP was measured and compared to the performance of k-NN applied on the original data. The underlying idea is that instances with the same class should get a similar feature representation, even though that class information is not used in the construction of the features.

Furthermore, totally random trees embedding [6] was also used in comparison. It was employed as an unsupervised transformation of the data, using a forest of Extremely Random Clustering trees (ERC) with a single random split candidate per node. In ERC the data are transformed using only the indices of the leaves of each tree. Similar to our approach, ERT was also chosen as the base estimator.

The number of trees used in the ensembles for all the compared methods was set to 300. At that number, the Gram matrix induced on the new features converged in the supervised setting [4]. The number of the features selected as splitting candidates (T_f) was set equal to the square root of the number of original features $(T_f = \sqrt{|M|})$. The variance over the feature set was computed as the sum of the variances over the individual features. All trees were unpruned, and the minimal number of instances a leaf has to cover was set equal to 3. As for k-NN, the 3 nearest neighbors were considered $(k = 3)$. Experiments selecting other numbers of nearest neighbors or splitting candidates (T_f) were also performed without showing a different trend. The evaluation was performed in a 10-fold cross validation (10 CV) scheme.

The evaluation measures that were employed were the common accuracy and the area under the receiver operating characteristic curve (AUROC). A ROC curve is defined as the true positive rate (TPR) against the false positive rate (FPR) at various thresholds. Alternatively, the true-positive rate is known as sensitivity and the false-positive rate as (1 - specificity).

As it is reflected in Tables 2 and 3 the proposed method $ERCP$ outperforms ERC in terms of $AUROC$. For each dataset, the best result is indicated with $*$. Furthermore, both tree-based ensemble methods succeed in generating a better feature representation set than the original one. More precisely, the average $AUROC$ results for $ERCP$ and ERC are 0.854 and 0.844, respectively. On the original set the average drops to 0.836. Further experiments were performed using different number of trees in the ensemble and different number of nearest neighbors. The obtained results, that are shown in Table 3, reaffirm the performance of the proposed approach. When it comes to accuracy the same behavior was witnessed as the average rates are 0.831, 0.827, and 0.824 for the $ERCP$, ERC, and the original set respectively.

In addition to k-NN, k-means was employed extending the evaluation of the proposed method to a clustering setting. Although there are many clustering algorithms, k-means was selected out of simplicity. The number of clusters was set equal to 2 as all the datasets contain 2 classes. The evaluation metric that was used was the adjusted Rand index [29], measuring the similarity between the

Table 2. *AUROC* measures for the compared approaches.

Data	Original	ERC	ERCP
Pima Indians diabetes	*0.767	0.726	0.731
Ecoli	*0.966	0.965	0.965
Glass identification	0.805	0.823	*0.871
Haberman's survival	0.629	0.609	*0.630
Ionosphere	0.897	0.937	*0.957
Iris	*1	*1	*1
Libras movement	0.753	*0.801	0.735
Robot execution failures (Lp5)	0.915	0.886	*0.968
Mammographic mass	0.791	*0.795	0.791
Sonar	0.718	0.713	*0.734
Spectf heart	0.707	0.748	*0.779
Statlog (vehicle)	0.981	*0.986	0.971
Breast cancer (orig.)	0.982	*0.983	*0.983
Breast cancer (diag.)	0.984	*0.985	0.977
Wine	0.970	*0.991	0.973
Breast cancer (prog.)	0.503	0.546	*0.590
Average	0.836	0.844	*0.854
Nb wins	3	7	*9
Average ranks	2.31	1.94	*1.75

Table 3. Average *AUROC* with different numbers of trees and nearest neighbors.

	ERC_{50}	$ERCP_{50}$	ERC_{100}	$ERCP_{100}$	ERC_{200}	$ERCP_{200}$	ERC_{400}	$ERCP_{400}$	Original
$k=2$	0.813	*0.830	0.827	*0.837	0.832	*0.840	0.840	*0.840	0.834
$k=4$	0.834	*0.852	0.837	*0.853	0.853	*0.855	0.850	*0.860	0.839
$k=5$	0.837	*0.857	0.842	*0.857	0.854	*0.855	0.853	*0.861	0.844
$k=6$	0.842	*0.858	0.844	*0.856	0.853	*0.856	0.854	*0.862	0.844
$k=8$	0.850	*0.859	0.847	*0.860	0.855	*0.857	0.857	*0.860	0.848

ground truth class assignments and the clustering algorithm assignments. The compared approaches correspond to different dimensional spaces, making the application of an evaluation metric based on the distances or the variances of the clusters difficult. Although the labels assigned to the samples by unsupervised clustering are without intrinsic meaning, the rational idea is that samples with the same ground truth are similar and therefore should be grouped together. As it is reflected in Table 4, the proposed method $ERCP$ outperforms the other comparing approaches for both $T_f = \sqrt{|M|}$ and $T_f = 1$. It is interesting to note that the best results in clustering (k-means) are obtained with $T_f = 1$

Table 4. Adjusted Rand index results for the compared approaches.

Data	Original	ERC	$ERCP_{T_f = \sqrt{|M|}}$	$ERCP_{(T_f = 1)}$
Pima Indians diabetes	⋆0.11	0.09	0.04	*0.15
Ecoli	⋆*0.62	0.58	0.58	0.58
Glass identification	0	0	0	0
Haberman's survival	0	0	0	0
Ionosphere	0.17	0.15	⋆0.18	*0.20
Iris	1	1	1	1
Libras movement	0	0	0	0
Robot execution failures (Lp5)	*−0.03	−0.07	⋆0.09	−0.07
Mammographic mass	⋆*0.36	0.30	0.31	0.32
Sonar	0	0	0	*0.01
Spectf heart	−0.1	*−0.07	⋆−0.04	*−0.07
Statlog (vehicle)	0.15	⋆*0.17	0.15	*0.17
Breast cancer (orig.)	0.84	0.82	⋆0.89	*0.89
Breast cancer (diag.)	0.65	⋆0.69	0.68	*0.73
Wine	0.01	⋆*0.11	0.02	*0.11
Breast cancer (prog.)	0.02	⋆*0.03	0.02	*0.03
Average	0.238	0.238	⋆0.244	*0.253
Nb wins⋆ (average ranks)	3(2.06)	4(2.04)	4(1.9)	-
Nb wins* (average ranks)	3(2.16)	4(2.22)	-	9(1.63)

(totally randomized tree-paths, as in ERC). The best results among the *ERC*, $ERCP_{T_f=\sqrt{|M|}}$, and the *original* features are reported with ⋆. The best results among the *ERC*, $ERCP_{T_f=1}$, and the *original* features are reported with *. It has to be mentioned that optimizing some parameters for each dataset was not part of the study performed here, even though it could possibly lead to better results.

In Figs. 3 and 4, a visualization of PPI and MN datasets (homogeneous networks) is displayed by projecting the data in a 2-dimensional (2D) space using PCA. Other linear or non-linear techniques such as the t-SNE [30] could have been used but the common PCA was chosen out of simplicity. As reflected in the Figs. 3 and 4, the generated data distribution after applying PCA to the original data fails to detect any underlying manifold and it is similar to a common random projection, especially for the MN dataset. In the case of *ERC*, two clusters seem to appear, however it is not clear where to dichotomize the data. Finally, the application of PCA to the *ERCP*-induced feature space leads to a more informative distribution and shows two clearly disconnected clusters. The two clusters have been color coded with colors blue and red, and the same coding scheme was applied in the other graphs. For the PPI dataset, a Gene Ontology Enrichment analysis was performed using YeastMine [31] in order to assign a biological interpretation to the obtained clusters. Using the complete set of proteins

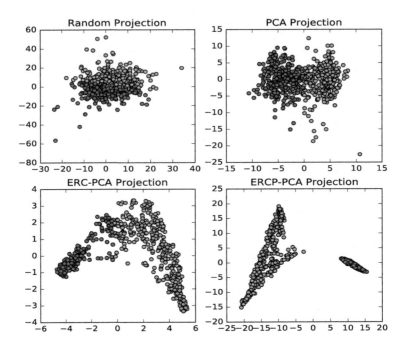

Fig. 3. MN network data projection. Upper left a totally random projection of the data is depicted. Upper right the PCA projection of the original data is shown. Down left the PCA projection of the *ERC* feature representation is displayed. Down right the PCA projection of the *ERPC* is displayed. (Color figure online)

as background, it turns out that the bigger cluster (red) is enriched with proteins localized in the nucleus (p = 3.26e-60), while the smaller cluster is enriched with cytoplasm cellular component annotations (p = 0.038). It is concluded that *ERCP* succeeds in providing a more informative feature representation for complex datasets.

Next, the experimental evaluation of the proposed interaction data mining scheme is presented. The global representation was constructed as described in Sect. 2.3. It consists of all the possible pairs of network nodes. For evaluation purposes, the known interactions or non-interactions between these nodes were coded as 1 and 0, respectively. They were used as ground truth without taking part in the learning process. Then, the performance of a k-NN algorithm applied to that global representation was measured. The global network representation based on the proposed approach that was described in Sect. 2.3 is referred to as MID-CT (Mining Interaction Data with Clustering Trees). The global representation based on the original features is coined as Global Network Representation (GNR) and a global representation based on the original features and PCA is coined as GNR-PCA. More specifically, in GNR-PCA only PCA is applied on the original features of each node-set. Here, the number of components that were kept was set equal to the square root of the original features ($\sqrt{|M|}$). In Table 5,

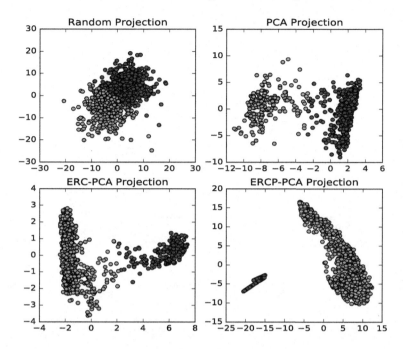

Fig. 4. PPI network data projection. Upper left a totally random projection of the data is depicted. Upper right the PCA projection of the original data is shown. Down left the PCA projection of the *ERC* feature representation is displayed. Down right the PCA projection of the *ERPC* is displayed. (Color figure online)

the accuracy results of k-NN for the first nearest neighbor (1-NN) as well as the sizes of the compared representations are shown. In Fig. 5, the AUROC values for different numbers of nearest neighbors are shown. As it is reflected, the MID-CT clearly outperforms the other approaches. It is also shown that the results are improved using high values of k in k-NN. To this end, it could be deducted that the representation yielded by our approach is characterized by more pure neighborhoods. Moreover, it has to be mentioned that MID-CT yields a computationally much more efficient representation than GNR as it reduces the size of the two interaction sets before the final construction of the global representation. This way, a global network representation of much less dimensions is obtained.

Table 5. Accuracy results (1-NN) for the compared approaches.

Dataset	Size of GNR	Size of MID-CT	GNR	GNR-PCA	MID-CT
DPI (drug-protein)	1215057 × 1536	1215057 × 56	0.7655	0.7757	*0.9180
SRN (genes-TF)	205773 × 11569	205773 × 140	0.5495	0.5510	*0.9293
ERN (genes-TF)	179256 × 890	179256 × 42	0.9415	0.9515	*0.9719

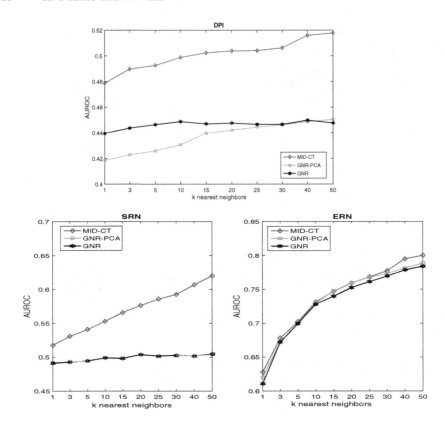

Fig. 5. AUROC results for different numbers of nearest neighbors.

4   Conclusions and Future Work

In this paper, we proposed an efficient feature representation framework based on decision tree ensembles for unsupervised learning tasks. In particular, we employed Extremely Randomized Trees in an unsupervised manner, by evaluating the quality of a split on the feature space, rather than the target space. By registering the tree nodes that are encountered by a given sample, a high-dimensional, very sparse feature space is obtained. The proposed approach is inductive and can handle complex data structures. Moreover, we proposed a new scheme based on the aforementioned approach for mining interaction data organized as heterogeneous networks. Finally, we empirically evaluated the proposed data representation using UCI datasets as well as more complex datasets representing interaction networks. The effectiveness of the approach was confirmed by showing improved performance when a mining algorithm or data visualisation step is applied on the obtained feature representation.

Possible topics for future research include the application of various machine learning algorithms to the generated feature representation or the development of an efficient weighing scheme, assigning a different weight to each tree-node of the ensemble. This way, the information contained in each generated feature will be distilled.

Acknowledgments. The authors acknowledge the Research Fund KU Leuven. They also want to thank Lieven Thorrez for input and feedback on the biological interpretation of the visualization results.

References

1. Lanckriet, G.R., Cristianini, N., Bartlett, P., Ghaoui, L.E., Jordan, M.I.: Learning the kernel matrix with semidefinite programming. J. Mach. Learn. Res. **5**, 27–72 (2004)
2. Shawe-Taylor, J., Cristianini, N.: Kernel Methods for Pattern Analysis. Cambridge University Press, Cambridge (2004)
3. Burges, C.J.: A tutorial on support vector machines for pattern recognition. Data Min. Knowl. Disc. **2**(2), 121–167 (1998)
4. Vens, C., Costa, F.: Random forest based feature induction. In: Proceedings of IEEE 11th International Conference on Data Mining (ICDM), pp. 744–753 (2011)
5. Breiman, L.: Random forests. Mach. Learn. **45**(1), 5–32 (2001)
6. Moosmann, F., Triggs, B., Jurie, F.: Fast discriminative visual codebooks using randomized clustering forests. In: Proceedings 20th Conference on Neural Information Processing Systems (NIPS), pp. 985–992 (2006)
7. Moosmann, F., Triggs, B., Jurie, F.: Randomized clustering forests for image classification. IEEE Trans. Pattern Anal. Mach. Intell. **30**(9), 1632–1646 (2008)
8. Zhang, M., Wu, L.: LIFT: multi-label learning with label-specific features. IEEE Trans. Pattern Anal. Mach. Intell. **37**(1), 107–120 (2015)
9. Blockeel, H., De Raedt, L.: Top-down induction of first-order logical decision trees. Artif. Intell. **101**(1), 285–297 (1998)
10. Blockeel, H., De Raedt, L., Ramon, J.: Top-down induction of clustering trees. In: Proceedings of the 15th International Conference on Machine Learning, pp. 55–63 (1998)
11. Geurts, P., Irrthum, A., Wehenkel, L.: Supervised learning with decision tree-based methods in computational and systems biology. Mol. BioSyst. **5**(12), 1593–1605 (2009)
12. Agichtein, E., Castillo, C., Donato, D., Gionis, A., Mishne, G.: Finding high-quality content in social media. In: Proceedings of the ACM International Conference on Web Search and Data Mining, pp. 183–194 (2008)
13. Kocev, D., Vens, C., Struyf, J., Džeroski, S.: Tree ensembles for predicting structured outputs. Pattern Recogn. **46**(3), 817–833 (2013)
14. Geurts, P., Ernst, D., Wehenkel, L.: Extremely randomized trees. Mach. Learn. **63**(1), 3–42 (2006)
15. Yan, S., Xu, D., Zhang, B., Zhang, H.J., Yang, Q., Lin, S.: Graph embedding and extensions: a general framework for dimensionality reduction. IEEE Trans. Pattern Anal. Mach. Intell. **29**(1), 40–51 (2007)
16. Van Der Maaten, L., Postma, E., Van den Herik, J.: Dimensionality reduction: a comparative review. J. Mach. Learn. Res. **10**, 66–71 (2009)
17. Schrynemackers, M., Wehenkel, L., Babu, M.M., Geurts, P.: Classifying pairs with trees for supervised biological network inference. Mol. BioSyst. **11**(8), 2116–2125 (2015)

18. Maetschke, S.R., Madhamshettiwar, P.B., Davis, M.J., Ragan, M.A.: Supervised, semi-supervised and unsupervised inference of gene regulatory networks. Briefings Bioinform. **15**(2), 195–211 (2014)

19. Stojanova, D., Ceci, M., Malerba, D., Dzeroski, S.: Using PPI network autocorrelation in hierarchical multi-label classification trees for gene function prediction. BMC Bioinform. **14**(1), 285 (2013)

20. Bleakley, K., Biau, G., Vert, J.P.: Supervised reconstruction of biological networks with local models. Bioinformatics **23**(13), i57–i65 (2007)

21. Vert, J.P., Qiu, J., Noble, W.S.: A new pairwise kernel for biological network inference with support vector machines. BMC Bioinform. **8**(10), 1 (2007)

22. Kocev, D., Ceci, M.: Ensembles of extremely randomized trees for multi-target regression. In: Japkowicz, N., Matwin, S. (eds.) DS 2015. LNCS, vol. 9356, pp. 86–100. Springer, Cham (2015). doi:10.1007/978-3-319-24282-8_9

23. Asuncion, A., Newman, D.: UCI machine learning repository. http://www.ics.uci.edu/mlearn/MLRepository.html

24. Yamanishi, Y., Vert, J.P., Kanehisa, M.: Supervised enzyme network inference from the integration of genomic data and chemical information. Bioinformatics **21**(Suppl. 1), i468–i477 (2005)

25. Von Mering, C., Krause, R., Snel, B., Cornell, M., Oliver, S.G., Fields, S., Bork, P.: Comparative assessment of large-scale data sets of protein-protein interactions. Nature **417**(6887), 399–403 (2002)

26. Faith, J.J., Hayete, B., Thaden, J.T., Mogno, I., Wierzbowski, J., Cottarel, G., Kasif, S., Collins, J.J., Gardner, T.S.: Large-scale mapping and validation of Escherichia coli transcriptional regulation from a compendium of expression profiles. PLoS Biol. **5**(1), e8 (2007)

27. MacIsaac, K.D., Wang, T., Gordon, D.B., Gifford, D.K., Stormo, G.D., Fraenkel, E.: An improved map of conserved regulatory sites for Saccharomyces cerevisiae. BMC Bioinform. **7**(1), 1 (2006)

28. Yamanishi, Y., Pauwels, E., Saigo, H., Stoven, V.: Extracting sets of chemical substructures and protein domains governing drug-target interactions. J. Chem. Inf. Model. **51**(5), 1183–1194 (2011)

29. Hubert, L., Arabie, P.: Comparing partitions. J. Classif. **2**(1), 193–218 (1985)

30. Van Der Maaten, L., Hinton, G.: Visualizing data using t-SNE. J. Mach. Learn. Res. **9**, 2579–2605 (2008)

31. Cherry, J.M., Hong, E.L., Amundsen, C., Balakrishnan, R., Binkley, G., Chan, E.T., Christie, K.R., Costanzo, M.C., Dwight, S.S., Engel, S.R., Fisk, D.G., Hirschman, J.E., Hitz, B.C., Karra, K., Krieger, C.J., Miyasato, S.R., Nash, R.S., Park, J., Skrzypek, M.S., Simison, M., Weng, S., Wong, E.D.: Saccharomyces Genome Database: the genomics resource of budding yeast. Nucleic Acids Res. **40**(Database issue), D700–D705 (2012)

Ranking to Learn:
Feature Ranking and Selection via Eigenvector Centrality

Giorgio Roffo[1,2]([⊠]) and Simone Melzi[2]

[1] School of Computing Science, University of Glasgow, Glasgow, UK
Giorgio.Roffo@glasgow.ac.uk
[2] Department of Computer Science, University of Verona, Verona, Italy
Simone.Melzi@univr.it

Abstract. In an era where accumulating data is easy and storing it inexpensive, feature selection plays a central role in helping to reduce the high-dimensionality of huge amounts of otherwise meaningless data. In this paper, we propose a graph-based method for feature selection that ranks features by identifying the most important ones into arbitrary set of cues. Mapping the problem on an affinity graph - where features are the nodes - the solution is given by assessing the importance of nodes through some indicators of centrality, in particular, the *Eigenvector Centrality (EC)*. The gist of EC is to estimate the importance of a feature as a function of the importance of its neighbors. Ranking central nodes individuates candidate features, which turn out to be effective from a classification point of view, as proved by a thoroughly experimental section. Our approach has been tested on 7 diverse datasets from recent literature (e.g., biological data and object recognition, among others), and compared against filter, embedded and wrappers methods. The results are remarkable in terms of accuracy, stability and low execution time.

Keywords: Feature selection · Ranking · High dimensionality · Data mining

1 Introduction

As data collection technologies advance and computer power grows, a torrent of data is generated in almost every field computers are used [5]. Because the volume, velocity, variety and complexity of datasets is continuously increasing, pattern recognition methodologies have become indispensable in order to extract useful information from huge amounts of otherwise meaningless data.

Feature Selection (FS) is one of the long existing methods that deals with these problems [14]. Its objective is to select a minimal subset of those attributes that allows a problem to be clearly defined. By choosing a minimal subset of features, irrelevant and redundant features are removed according to some reasonable criteria so that the original task can be achieved equally well, if not better.

© Springer International Publishing AG 2017
A. Appice et al. (Eds.): NFMCP 2016, LNAI 10312, pp. 19–35, 2017.
DOI: 10.1007/978-3-319-61461-8_2

FS techniques can be partitioned into three classes [14]: *wrappers* (see Fig. 1), which use classifiers to score a given subset of features; *embedded* methods (in Fig. 3), which inject the selection process into the learning of the classifier; and *filter* methods (see Fig. 2), which analyze intrinsic properties of data, ignoring the classifier. Filters are also (relatively) robust against overfitting.

Most of these methods can perform two operations, *ranking* and *subset selection*: in the former, the importance of each individual feature is evaluated, usually by neglecting potential interactions among the elements of the joint set [8]; in the latter, the final subset of features to be selected is provided. In some cases, these two operations are performed sequentially (first the ranking, then the selection) [7,12,17,24,35]; in other cases, only the selection is carried out [13]. Usually, the subset selection is supervised, while in the ranking case, methods can be supervised or not. FS is *NP-hard* [14]; if there are n features in total, the goal is to select the optimal subset of $m \ll n$, by evaluating $\binom{n}{m}$ combinations; therefore, suboptimal search strategies are considered (see Sect. 2 for an overview). With the filters, features are first considered individually, ranked, and then a subset is extracted, some examples are Mutual Information [35], Relief-F [24], Inf-FS [30,31] unsupervised and not [26], and mRMR [27]. Conversely, with wrapper and embedded methods, subsets of features are sampled, evaluated, and finally kept as the final output, for instance, FSV[7,12] and SVM-RFE [17].

In this work, we propose a novel graph-based feature selection algorithm that ranks features according to a graph centrality measure (Eigenvector centrality [6]). The main idea behind the method is to map the problem on an affinity graph, and to model pairwise relationships among feature distributions by weighting the edges connecting them.

The novelty of the proposed method in terms of the state of the art is that it assigns a score of "importance" to each feature by taking into account all the other features mapped as nodes on the graph, bypassing the combinatorial problem in a methodologically sound fashion. Indeed, eigenvector centrality differs from other measurements (e.g., degree centrality) since a node - feature - receiving many links does not necessarily have a high eigenvector centrality. The reason is that not all nodes are equivalent, some are more relevant than others, and, reasonably, endorsements from important nodes count more (see Sect. 3.2). Noteworthy, another important contribution of this work is the scalability of the method. Indeed, centrality measurements can be implemented using the Map Reduce paradigm [20,23,34], which makes the algorithm prone to a possible distributed version [29].

Our approach is extensively tested on 7 benchmarks of cancer classification and prediction on genetic data (*Colon* [2], *Prostate* [11], *Leukemia* [11], *Lymphoma* [11]), handwritten recognition (GINA [1]), generic feature selection datasets (MADELON [15]), and object recognition (PASCAL VOC 2007 [9]). We compare the proposed method on these data, while comparing it against seven efficient approaches under different conditions (number of features selected and number of training samples considered), overcoming all of them in terms of ranking stability or classification accuracy.

Finally, we provide an open and portable library of feature selection algorithms, integrating the methods with uniform input and output formats to facilitate large scale performance evaluation. The *Feature Selection Library* (FSLib Matlab Toolbox[1]) and interfaces are fully documented. The library integrates directly with MATLAB, a popular language for machine learning and pattern recognition research.

The rest of the paper is organized as follows. A brief overview of the related literature is given in Sect. 2, mostly focusing on the comparative approaches we consider in this work. Our feature selection algorithm is described in Sect. 3. Graph construction and weighting are presented in Sects. 3.1 and 3.2 respectively, while the employed Eigenvector centrality is discussed in Sect. 3.3. Section 4 contains the experimental evaluations and results. Finally, conclusions are provided in Sect. 6.

2 Related Work

Since the mid-1990s, few domains explored used more than 50 features. The situation has changed considerably in the past few years and most papers explore domains with hundreds to tens of thousands of features. New approaches are proposed to address these challenging tasks involving many irrelevant and redundant cues and often comparably few training examples. Among the most used FS strategies, *Relief-F* [24] is an iterative, randomized, and supervised approach that estimates the quality of the features according to how well their values differentiate data samples that are near to each other; it does not discriminate among redundant features (i.e., may fail to select the most useful features), and performance decreases with few data. Similar problems affect SVM-RFE (RFE) [17], which is a wrapper method (see Fig. 1) that selects features in a sequential, backward elimination manner, ranking high a feature if it strongly separates the samples by means of a linear SVM.

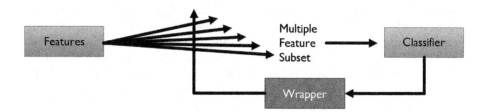

Fig. 1. Wrapper models involve optimizing a predictor as part of the selection process. They tend to give better results but filter methods are usually computationally less expensive than wrappers.

---

[1] The FSLib is publicly available on File Exchange - MATLAB Central at: https://it. mathworks.com/matlabcentral/fileexchange/56937-feature-selection-library.

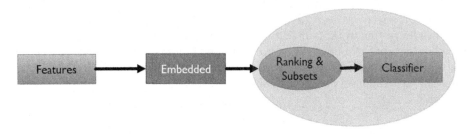

Fig. 2. Filter methods: the selection of features is independent of the classifier used. They rely on the general characteristics of the training data to select features with independence of any predictor.

Fig. 3. In embedded methods the learning part and the feature selection part can not be separated.

Batti [4] has developed the Mutual Information-Based Feature Selection (MIFS) criterion, where the features are selected in a greedy manner. Given a set of existing selected features, at each step it locates the feature x_i that maximizes the relevance to the class. The selection is regulated by a proportional term β that measures the overlap information between the candidate feature and existing features. In [36] the authors proposed a *graph-based* filter approach to feature selection, that constructs a graph in which each node corresponds to each feature, and each edge has a weight corresponding to mutual information (MI) between features connected by that edge. This method performs dominant set clustering to select a highly coherent set of features and then it selects features based on the multidimensional interaction information (MII). Another effective yet fast filter method is the *Fisher* method [13], it computes a score for a feature as the ratio of interclass separation and intraclass variance, where features are evaluated independently, and the final feature selection occurs by aggregating the m top ranked ones. Other widely used filters are based on mutual information, dubbed *MI* here [35], which considers as a selection criterion the mutual information between the distribution of the values of a given feature and the membership to a particular class. Mutual information provides a principled way of measuring the mutual dependence of two variables, and has been used by a number of researchers to develop information theoretic feature selection criteria. Even in the last case, features are evaluated independently, and the final feature selection occurs by aggregating the m top ranked ones. Maximum-Relevance Minimum-Redundancy criterion (MRMR) [27] is an efficient incremental search algorithm. Relevance scores are assigned by maximizing the joint mutual information between the class variables and the subset of selected features. The computation of the information between high-dimensional vectors is impractical, as the time required becomes prohibitive. To face this problem the mRMR propose to estimate the mutual information for continuous variables using Parzen Gaussian windows. This estimate is based on a heuristic framework to minimize redundancy and uses a series of intuitive measures of relevance and redundancy to select features. Note, it is equivalent to MIFS with $\beta = \frac{1}{n-1}$, where n is the number of features. Selecting features in unsupervised learning scenarios is a much harder problem, due to the absence of class labels that would guide the search for relevant information. In this scenario, we compare our approach against the recent unsupervised graph-based filter dubbed Inf-FS [31]. In the Inf-FS formulation, each feature is a node in the graph, a path is a selection of features, and the higher the centrality score, the most important (or most different) the feature. It assigns a score of "importance" to each feature by taking into account all the possible feature subsets as paths on a graph. Another unsupervised method is the Laplacian Score (LS) [19], where the importance of a feature is evaluated by its power of locality preserving. In order to model the local geometric structure, this method constructs a nearest neighbor graph. LS algorithm seeks those features that respect this graph structure. Finally, for the embedded method (see Fig. 3), we include the *feature selection via concave minimization (FSV)* [7], where the selection process is injected *into* the training of an SVM by a linear programming technique.

3 Proposed Method

3.1 Building the Graph

Given a set of features $X = \{x^{(1)}, \ldots, x^{(n)}\}$ we build an undirected graph $G = (V, E)$; where V is the set of vertices corresponding, one by one, to each variable x. E codifies (weighted) edges among features. Let the adjacency matrix A associated with G define the nature of the weighted edges: each element a_{ij} of A, $1 \leq i, j \leq n$, represents a pairwise potential term. Potentials can be represented as a binary function $\varphi(x^{(i)}, x^{(j)})$ of the nodes $x^{(k)}$ such as:

$$a_{ij} = \varphi(x^{(i)}, x^{(j)}). \tag{1}$$

The graph can be weighted according to different heuristics, therefore the function φ can be handcrafted or automatically learned from data.

3.2 φ-Design

The design of the φ function is a crucial operation. In this work, we weight the graph according to good reasonable criteria, related to class separation, so as to address the classification problem. In other words, we want to rank features according to how well they discriminate between two classes. Hence, we draw upon best-practice in FS and propose an ensemble of two different measures capturing both relevance (supervised) and redundancy (unsupervised) proposing a kernelized-based adjacency matrix. Before continuing with the discussion, note that each feature distribution $x^{(i)}$ is normalized so as to sum to 1.

Firstly, we apply the Fisher criterion:

$$f_i = \frac{|\mu_{i,1} - \mu_{i,2}|^2}{\sigma_{i,1}^2 + \sigma_{i,2}^2},$$

where $\mu_{i,C}$ and $\sigma_{i,C}$ are the mean and standard deviation, respectively, assumed by the i-th feature when considering the samples of the C-th class. The higher f_i, the more discriminative the i-th feature. However, a natural generalization of this score into a multi-class framework is given by

$$f_i = \frac{\sum_{c=1}^{\mathbf{C}} (\mu_{i,c} - \mu_i)^2}{\sigma_i^2}$$

where μ_i and σ_i denote the mean and standard deviation of the whole data set corresponding to the i-th feature (i.e., $\sigma_i^2 = \sum_{c=1}^{\mathbf{C}} (\sigma_{i,c})^2$).

Because we are given class labels, it is natural that we want to keep only the features that are related to or lead to these classes. Therefore, we use mutual information to obtain a good feature ranking that score high features highly predictive of the class.

$$m_i = \sum_{y \in Y} \sum_{z \in x^{(i)}} p(z,y) log\left(\frac{p(z,y)}{p(z)p(y)}\right),$$

where Y is the set of class labels, and $p(\cdot, \cdot)$ the joint probability distribution.

A kernel k is then obtained by the matrix product

$$k = (f \cdot m^\top),$$

where f and m are $n \times 1$ column vectors normalized in the range 0 to 1, and k results in a $n \times n$ matrix.

To boost the performance, we introduce a second feature-evaluation metric based on standard deviation [17] – capturing the amount of variation or dispersion of features from average – as follows:

$$\Sigma(i,j) = max\left(\sigma^{(i)}, \sigma^{(j)}\right),$$

where σ being the standard deviation over the samples of x, and Σ turns out to be a $n \times n$ matrix with values $\in [0,1]$.

Finally, the adjacency matrix A of the graph G is given by

$$A = \alpha k + (1 - \alpha)\Sigma, \tag{2}$$

where α is a loading coefficient $\in [0,1]$. The generic entry a_{ij} accounts for how much discriminative are the feature i and j when they are jointly considered; at the same time, a_{ij} can be considered as a weight of the edge connecting the nodes i and j of a graph, where the i-th node models the i-th feature distribution (we report the sketch of our method in Algorithm 1).

Algorithm 1. Eigenvector Centrality Feature Selection (EC-FS)

Input: $X = \{x^{(1)}, ..., x^{(n)}\}$, $Y = \{y^{(1)}, ..., y^{(n)}\}$
Output: v_0 ranking scores for each feature
  - Building the graph
  C_1 positive class, C_2 negative class
  for $i = 1 : n$ **do**
    Compute $\mu_{i,1}$, $\mu_{i,2}$, $\sigma_{i,1}$, and $\sigma_{i,2}$
    Fisher score: $f(i) = \frac{(\mu_{i,1} - \mu_{i,2})^2}{\sigma_{i,1}^2 + \sigma_{i,2}^2}$
    Mutual Information: $m(i) = \sum_{y \in Y} \sum_{z \in x^{(i)}} p(z, y) log \left(\frac{p(z,y)}{p(z)p(y)} \right)$
  end for
  for $i = 1 : n$ **do**
    for $j = 1 : n$ **do**
      $k(i, j) = f(i)m(j)$,
      $\Sigma(i, j) = max \left(\sigma^{(i)}, \sigma^{(j)} \right)$,
      $A(i, j) = \alpha k(i, j) + (1 - \alpha)\Sigma(i, j)$
    end for
  end for
  - Ranking
  Compute eigenvalues $\{\Lambda\}$ and eigenvectors $\{V\}$ of A
  $\lambda_0 = \max_{\lambda \in \Lambda}(abs(\lambda))$
  return v_0 the eigenvector associated to λ_0

3.3   Eigenvector Centrality

From a graph theory perspective identifying the most important nodes corresponds to individuate some indicators of centrality within a graph (e.g., the relative importance of nodes). A first way used in graph theory is to study accessibility of nodes, see [10,28] for example. The idea is to compute A^l for some suitably large l (often the diameter of the graph), and then use the row sums of its entries as a measure of accessibility (i.e. $scores(i) = [A^l \mathbf{e}]_i$, where $\mathbf{e}$ is a vector with all entries equal to 1). The accessibility index of node i would thus be the sum of the entries in the i-th row of A^l, and this is the total number of paths of length l (allowing stopovers) from node i to all nodes in the graph. One problem with this method is that the integer l seems arbitrary. However, as we count longer and longer paths, this measure of accessibility converges to a index known as eigenvector centrality measure (EC) [6].

The basic idea behind the EC is to calculate v_0 the eigenvector of A associated to the largest eigenvalue. Its values are representative of how strongly each node is connected to the other nodes. Since the limit of A^l as l approaches a large positive number L converges to v_0,

$$\lim_{l \to L} [A^l \mathbf{e}] = v_0, \tag{3}$$

the EC index makes the estimation of indicators of centrality free of manual tuning over l, and computationally efficient.

Let us consider a vector, for example $\mathbf{e}$, that is *not* orthogonal to the principal vector v_0 of A. It is always possible to decompose $\mathbf{e}$ using the eigenvectors as basis with a coefficient $\beta_0 \neq 0$ for v_0. Hence:

$$\mathbf{e} = \beta_0 v_0 + \beta_1 v_1 + \ldots + \beta_n v_n, \quad (\beta_0 \neq 0). \tag{4}$$

Then

$$Ae = A(\beta_0 v_0 + \beta_1 v_1 + \ldots + \beta_n v_n) = \beta_0 A v_0 + \beta_1 A v_1 + \ldots + \beta_n A v_n$$
$$= \beta_0 \lambda_0 v_0 + \beta_1 \lambda_1 v_1 + \ldots + \beta_n \lambda_n v_n. \tag{5}$$

So in the same way:

$$A^l \mathbf{e} = A^l(\beta_0 v_0 + \beta_1 v_1 + \ldots + \beta_n v_n) = \beta_0 A^l v_0 + \beta_1 A^l v_1 + \ldots + \beta_n A^l v_n$$
$$= \beta_0 \lambda_0^l v_0 + \beta_1 \lambda_1^l v_1 + \ldots + \beta_n \lambda_n^l v_n, \quad (\beta_0 \neq 0). \tag{6}$$

Finally we divide by the constant $\lambda_0^l \neq 0$ (see Perron-Frobenius theorem [25]),

$$\frac{A^l \mathbf{e}}{\lambda_0^l} = \beta_0 v_0 + \frac{\lambda_1^l \beta_1 v_1}{\lambda_0^l} + \ldots + \frac{\lambda_n^l \beta_n v_n}{\lambda_0^l}, \quad (\beta_0 \neq 0). \tag{7}$$

The limit of $\frac{A^l \mathbf{e}}{\lambda_0^l}$ as l approaches infinity equals $\beta_0 v_0$ since $\lim_{l \to \infty} \frac{\lambda_i^l}{\lambda_0^l} = 0, \forall l > 0$. What we see here is that as we let l increase, the ratio of the components of $A^l \mathbf{e}$ converges to v_0. Therefore, marginalizing over the columns of A^l, with a sufficiently large l, corresponds to calculate the principal eigenvector of matrix A [6]. Figure 4 illustrates a toy example of three random planar graphs. Graphs are made of 700 nodes and they are weighted by the Euclidean distance between each pair of points. In the example, high scoring nodes are those ones farther from the mean (i.e., the distance is conceived as quantity to maximize), the peculiarity of the eigenvector centrality is that a node is important if it is linked to by other important nodes (higher scores).

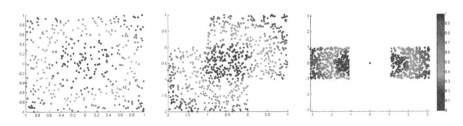

Fig. 4. Eigenvector centrality plots for three random planar graphs. On the left, a simple Gaussian distribution where central nodes are at the peripheral part of the distribution as expected. The central and right plots, some more complicated distributions, a node receiving many links does not necessarily have a high eigenvector centrality. Best viewed in color.

To the aim of this work, the use of eigenvector centrality allows to individuate candidate features, which turn out to be effective from a classification point of view, since indicators of centrality characterize the *global* (as opposed to local) prominence of a feature in the graph. Summarizing, the gist of eigenvector centrality is to compute the centrality of a node as a function of the centralities of its neighbors.

4   Experiments and Results

4.1   Datasets and Comparative Approaches

The datasets are chosen for letting the proposed method deal with diverse FS scenarios, as shown on Table 1. In the details, we consider the problems of dealing with few training samples and many features (*few train* in the table), unbalanced classes (*unbalanced*), or classes that severely overlap (*overlap*), or whose samples are noisy (*noise*) due to: (a) complex scenes where the object to be classified is located (as in the VOC series) or (b) many outliers (as in the genetic datasets, where samples are often *contaminated*, that is, artefacts are injected into the data during the creation of the samples). Lastly we consider the *shift* problem, where the samples used for the test are not congruent (coming from the same experimental conditions) with the training data.

Table 1. This table reports several attributes of the datasets used. The abbreviation *n.s.* stands for *not specified* (for example, in the object recognition datasets, the features are not given in advance).

Name	# samples	# classes	# feat.	*few train*	*unbal. (+/−)*	*overlap*	*noise*	*shift*
GINA [1]	3153	2	970			X		
MADELON [16]	4.4K	2	500			X		
Colon [2]	62	2	2K	X	(40/22)		X	
Lymphoma [11]	45	2	4026	X	(23/22)			
Prostate [33]	102	2	6034	X	(50/52)			
Leukemia [11]	72	2	7129	X	(47/25)		X	X
VOC 2007 [9]	10K	20	n.s		X		X	

Table 2 lists the methods in comparison, whose details can be found in Sect. 2. Here we just note their *type*, that is, f = filters, w = wrappers, e = embedded methods, and their *class*, that is, s = supervised or u = unsupervised (using or not using the labels associated with the training samples in the ranking operation). Additionally, we report their computational complexity (if it is documented in the literature). The computational complexity of our approach is $O(Tn + n^2)$.

The term Tn is due to the computation of the mean values among the T samples of every feature (n). The n^2 concerns the construction of the matrix A.

Table 2. List of the FS approaches considered in the experiments, specified according to their *Type*, class (*Cl.*), and complexity (*Compl.*). As for the complexity, T is the number of samples, n is the number of initial features, K is a multiplicative constant, i is the number of iterations in the case of iterative algorithms, and C is the number of classes. N/A indicates that the computational complexity is not specified in the reference paper.

Acronym	Type	Cl	Compl.
Fisher [13]	f	s	$\mathcal{O}(Tn)$
FSV [7,12]	e	s	N/A
Inf-FS [31]	f	u	$\mathcal{O}(n^{2.37}(1+T))$
MI [35]	f	s	$\sim\mathcal{O}(n^2 T^2)$
LS [19]	f	u	N/A
Relief-F [24]	f	s	$\mathcal{O}(iTnC)$
RFE [17]	w/e	s	$\mathcal{O}(T^2 n log_2 n)$
Ours	f	s	$\mathcal{O}(Tn + n^2)$

As for the computation of the leading eigenvector, it costs $O(m^2 n)$, where m is a number much smaller than n that is selected within the algorithm [22]. In the case that the algorithm can not be executed on a single computer, we refer the reader to [20,23,29,34] for distributed algorithms.

4.2   Exp. 1: Deep Representation (CNN) with Pre-training

This section proposes a set of tests on the PASCAL VOC-2007 [9] dataset. In object recognition VOC-2007 is a suitable tool for testing models, therefore, we use it as reference benchmark to assess the strengths and weaknesses of using our approach regarding the classification task. For this reason, we compare our approach against 8 state-of-the-art FS methods reported in Table 2. This experiment considers as features the cues extracted with a deep convolutional neural network architecture (CNN). We selected the pre-trained model called very deep ConvNets [32], which performs favorably to the state of the art for classification and detection in the ImageNet Large-Scale Visual Recognition Challenge 2014 (ILSVRC). We use the 4,096-dimension activations of the last layer as image descriptors (i.e., 4,096 features in total). The VOC-2007 edition contains about 10,000 images split into train, validation, and test sets, and labeled with twenty object classes. A one-vs-rest SVM classifier for each class is learnt (where cross-validation is used to find the best parameter C and α mixing coefficient in Eq. 2 on the training data) and evaluated independently and the performance is measured as mean Average Precision (mAP) across all classes.

Table 3 serves to analyze and empirically clarify how well important features are ranked high by several FS algorithms. The amount of features used for the two experiments is very low: $\approx 3\%$ and $\approx 6\%$ of the total. The results are significant: our method achieved the best performance in terms of mean average

Table 3. Varying the cardinality of the selected features. The image classification results achieved in terms of average precision (AP) scores while selecting the first 128 (3%) and 256 (6%) features from the total 4,096.

	PASCAL VOC 2007															
	First 128/4096 Features Selected								First 256/4096 Features Selected							
	Fisher	FSV	Inf-FS	LS	MI	ReliefF	RFE	Ours	Fisher	FSV	Inf-FS	LS	MI	ReliefF	RFE	Ours
✈	52.43	87.90	88.96	**89.37**	12.84	57.20	86.42	88.09	82.65	90.22	**91.16**	90.94	73.51	81.67	88.17	90.79
🚲	13.49	80.74	80.43	80.56	13.49	49.10	**82.14**	80.94	83.21	80.07	83.36	84.21	75.04	71.27	83.30	**84.72**
🐦	85.46	86.77	87.04	86.96	80.91	75.42	83.16	**88.74**	89.14	86.15	88.88	**89.31**	85.48	83.54	86.12	89.15
⛴	79.04	83.58	85.31	83.51	61.50	63.75	78.55	**86.90**	87.05	80.68	87.24	**87.84**	75.25	73.30	86.13	87.42
🍾	46.61	39.80	44.83	**49.36**	35.39	18.33	46.24	47.37	52.54	49.00	52.65	49.44	48.94	35.67	47.28	**53.20**
🚌	12.29	72.89	76.69	**76.98**	12.29	31.54	74.68	76.27	77.32	78.69	79.23	79.97	59.23	63.83	79.38	**80.57**
🚗	82.09	78.61	85.78	85.82	63.58	74.95	83.94	**85.92**	85.86	84.01	86.74	**87.06**	85.27	82.76	85.61	86.56
🐱	75.29	82.25	**83.34**	81.81	40.96	66.95	81.02	83.29	83.46	83.49	**85.61**	84.98	79.16	76.78	84.50	85.57
🪑	54.81	52.37	58.62	60.07	16.95	29.07	59.84	**60.57**	63.14	62.54	63.93	64.23	63.20	48.19	62.16	**64.53**
🐄	47.98	61.68	59.23	**65.50**	11.42	11.42	62.96	60.55	66.51	70.18	67.96	**71.54**	22.96	51.28	64.20	69.71
🪑	49.68	63.50	67.69	63.86	12.62	12.62	67.05	**67.70**	68.42	69.27	**71.78**	71.01	65.77	52.24	71.43	70.95
🐕	81.06	80.57	83.16	**83.21**	70.70	68.12	80.07	83.00	84.24	84.15	85.08	**85.20**	82.03	74.85	83.52	**85.20**
🐎	74.91	**83.33**	81.23	81.75	14.13	63.06	81.55	82.79	**85.68**	83.13	85.28	85.41	71.36	75.53	83.47	85.28
🏍	13.18	71.42	81.32	80.24	13.18	34.43	76.57	**82.20**	**84.29**	81.16	84.20	83.81	81.01	70.68	82.97	84.12
🧍	**91.33**	90.03	89.10	89.33	91.08	88.85	89.03	91.27	91.95	89.99	90.65	90.64	91.77	90.38	90.64	**91.99**
🪴	47.89	39.40	45.38	47.94	13.23	13.30	48.61	**49.05**	54.94	47.95	53.86	54.31	48.98	34.74	50.18	**55.88**
🐑	10.87	68.82	73.35	**74.05**	10.87	10.87	66.86	73.80	73.43	75.84	79.01	**81.57**	10.87	11.73	75.47	78.85
🛋	45.87	56.08	58.94	58.92	13.30	13.31	**62.06**	61.32	66.46	59.77	63.07	63.92	58.78	44.74	**66.68**	64.86
🚆	63.51	88.52	91.42	**91.48**	58.62	73.32	88.46	91.30	84.05	90.61	**93.21**	93.16	81.33	82.93	90.24	92.31
📺	64.29	65.61	66.79	62.99	47.25	24.96	67.10	**67.30**	71.44	69.19	70.56	70.75	71.39	55.59	**73.17**	72.49
mAP	54.60	71.69	74.43	74.69	34.72	44.03	73.32	**75.42**	76.79	75.80	78.17	78.47	66.57	63.09	76.73	**78.71**

precision (mAP) followed by the unsupervised filter methods LS and Inf-FS. As for the methods in comparison, one can observe the high variability in classification accuracy; indeed, results show that our method is robust to classes (i.e., by changing the testing class its performance is always comparable with the top scoring method).

4.3 Exp. 2: Testing on Microarray Databases

In application fields like biology is inconceivable to devise an analysis procedure which does not comprise a FS step. A clear example can be found in the analysis of expression microarray data, where the expression level of thousands of genes is simultaneously measured. Within this scenario, we tested the proposed approach on four well-known microarray benchmark datasets for two-class problems. Results are reported in Table 4. The testing protocol adopted in this experiment consists in splitting the dataset up to 2/3 for training and 1/3 for testing. In order to have a fair evaluation, the feature ranking has been calculated using only the training samples, and then applied to the testing samples. The classification is performed using a linear SVM. For setting the best parameters (C of the linear SVM, and α mixing coefficient) we used a 5-fold cross validation on the training data. This procedure is repeated several times and results are averaged over the trials. Results are reported in terms of the Receiver Operating Characteristic or ROC curves. A widely used measurement that summarizes the ROC curve is the Area Under the ROC Curve (AUC) [3] which is useful for comparing algorithms independently of application. Hence, classification results for the datasets used show that the proposed approach produces superior results in all the cases. The overall performance indicates that our approach is more robust than the others, by changing the data it still produces high quality rankings.

Table 4. The tables show results obtained on the expression microarray scenario. Tests have been repeated 100 times, and the means of the computed AUCs are reported for each dataset.

Microarray databases												
	Colon						Leukemia					
	# Features						# Features					
Method	50	100	150	200	Average	Time	50	100	150	200	Average	Time
Fisher-S	91.25	88.44	89.38	87.81	89.22	0.02	99.33	99.78	99.78	99.78	99.66	0.01
FSV	85.00	88.12	89.38	89.69	88.04	0.18	98.22	98.44	99.11	99.33	98.77	0.37
Inf-FS	88.99	89.41	89.32	89.01	89.18	0.91	99.91	**99.92**	**99.97**	**99.98**	**99.95**	5.49
LS	90.31	89.06	89.38	90.00	89.68	0.03	98.67	99.33	99.56	99.56	99.28	0.07
MI	89.38	90.31	90.63	**90.94**	90.31	0.31	99.33	99.33	99.56	99.33	98.38	0.21
ReliefF	80.94	84.38	85.94	87.50	84.69	0.52	99.56	99.78	99.78	99.78	99.72	1.09
RFE	89.06	85.00	86.88	85.62	86.64	0.18	**100**	99.78	99.56	99.78	99.78	0.14
EC-FS	**91.40**	**91.10**	**91.11**	90.63	**91.06**	0.45	99.92	**99.92**	99.77	99.85	99.86	1.50
	Lymphoma						Prostate					
	# Features						# Features					
Method	50	100	150	200	Average	Time	50	100	150	200	Average	Time
Fisher-S	98.75	98.38	98.38	100	98.87	0.01	96.10	96.20	96.30	97.30	96.47	0.02
FSV	98.22	98.44	99.11	99.33	98.77	0.18	96.70	96.70	96.50	96.30	96.55	0.63
Inf-FS	98.12	98.75	98.75	99.38	98.75	7.61	**96.80**	96.90	**97.10**	96.70	96.87	26.85
LS	90.00	96.88	99.38	98.75	96.25	0.04	85.80	94.60	96.90	97.00	93.57	0.24
MI	97.50	98.75	99.38	99.38	98.75	0.59	96.00	**96.90**	96.00	96.20	96.27	1.01
ReliefF	96.80	97.00	98.80	98.80	97.85	0.74	92.72	93.46	93.62	93.85	93.41	2.68
RFE	96.00	98.00	98.80	99.00	97.95	0.02	93.40	96.40	**97.10**	96.32	95.80	0.3
EC-FS	**99.40**	**99.20**	**99.60**	**99.20**	**99.20**	1.50	96.28	**96.90**	96.80	**98.10**	**97.02**	2.81

The quality of a feature subset is measured by an estimate of the classification accuracy of a chosen classifier trained on the candidate subset. Stability of the ranking is an important aspect when the task is knowledge discovery. The rationale behind this fact is that the estimate of the quality of the candidate subsets usually depends on many the training/testing split of the data. Therefore different sequences of features may be returned from repeated runs of FS approaches. In such a case, it is important to define if these numerous different subsets of features have approximately equal quality, otherwise presenting the user with only one subset may be misleading. We assessed the stability of the selected features using the Kuncheva index [21]. This stability measure represents the similarity between the set of rankings generated over the different splits of the dataset. The similarity between sequences of size N can be seen as the number of elements n they have in common (i.e. the size of their intersection). The Kuncheva index takes values in $[-1, 1]$, and the higher its value, the larger the number of commonly selected features in both sequences. The index is shown in Fig. 5, comparing our approach and the other methods. A valid alternative is the stability index based on Jensen-Shannon Divergence D_{JS}, proposed by [18], with a $[0, 1]$ range, where 0 indicates completely random rankings and 1 means stable rankings. Unlike Kuncheva measure, this metric is suitable for different algorithm outcomes: partial sublists (top-k lists) as well as the least studied partial ranked lists. Since in our case we work with full ranked lists, because all the feature selection algorithms considered in this study produce permutations of the original set of features, we preferred the widely used Kuncheva index. The proposed method shows, in most of the cases, a high stability whereas the highest performance is achieved.

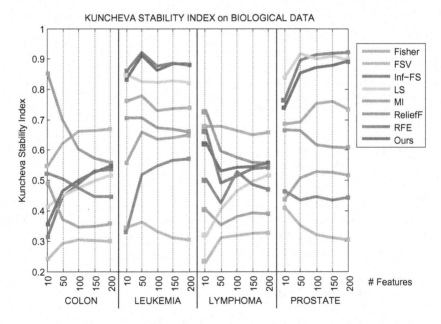

Fig. 5. The Kuncheva stability indices for each method in comparison are presented. The figure reports the stability while varying the cardinality of the selected features from 10 to 200 on different benchmarks.

4.4   Exp. 2: Other Benchmarks

GINA has sparse input variables consisting of 970 features. It is a balanced data set with 49.2% instances belonging to the positive class. Results obtained on GINA indicate that the proposed approach overcomes the methods in comparison, and select the most useful features from a data set with high-complexity and dimensionality. MADELON is an artificial dataset, which was part of the NIPS 2003 feature selection challenge. It represents a two-class classification problem with continuous input variables. The difficulty is that the problem is multivariate and highly non-linear. Results are reported in Table 5. This gives a proof about the classification performance of our approach that is attained on the test sets of GINA and MADELON.

Table 5. Varying the cardinality of the selected features. (ROC) AUC (%) on different datasets by SVM classification. Performance obtained with the first 50, 100, 150, and 200 features.

FS challenge datasets												
	GINA - handwritten recognition						MADELON - artificial data					
	# Features						# Features					
Method	50	100	150	200	Average	Time	50	100	150	200	Average	Time
Fisher-S	89.8	89.4	90.2	**90.4**	89.9	0.05	61.9	63.0	62.3	64.0	62.5	0.02
FSV	81.9	83.7	82.0	83.6	82.7	138	59.9	60.6	61.0	61.0	60.7	732
Inf-FS	89.0	88.7	89.1	89.0	88.9	41	62.6	**63.8**	**65.4**	60.8	63.2	0.04
LS	82.2	82.4	83.4	83.2	82.7	1.30	62.8	62.9	63.3	64.7	63.4	8.13
MI	89.3	89.7	89.8	90.1	89.6	1.13	63.0	63.7	63.5	64.7	63.6	0.4
ReliefF	77.9	76.3	77.3	76.9	77.2	0.12	62.9	63.1	63.2	**64.9**	63.5	10.41
RFE	82.2	82.4	83.4	83.2	82.7	6.60	55.0	61.2	57.1	60.2	56.5	50163
EC-FS	**90.9**	**90.3**	**90.4**	89.5	**90.3**	1.56	**63.6**	**63.8**	63.7	63.3	**63.7**	0.57

FS techniques definitely represent an important class of preprocessing tools, by eliminating uninformative features and strongly reducing the dimension of the problem space, it allows to achieve high performance, useful for practical purposes in those domains where high speed is required.

5   Reliability and Validity

In order to assess if the difference in performance is statistically significant, t-tests have been used for comparing the accuracies. Statistical tests are used to determine if the accuracies obtained with the proposed approach are significantly different from the others (whereas both the distribution of values were normal). The test for assessing whether the data come from normal distributions with unknown, but equal, variances is the *Lilliefors* test. Results have been obtained by comparing the results produced by each method over 100 trials (at each trial corresponds a different split of the data). Given the two distributions x_p of the proposed method and x_c of the current competitor, of size 1×100, a *two-sample t-test* has been applied obtaining a test decision for the null hypothesis that the data in vectors x_p and x_c comes from independent random samples from normal distributions with equal means and equal but unknown variances. Results (highlighted in Tables 4 and 5) show a statistical significant effect in performance (p-value < 0.05, Lilliefors test $H = 0$).

6   Conclusion

In this paper we present the idea of solving feature selection via the Eigenvector centrality measure. We design a graph – where features are the nodes – weighted by a kernelized adjacency matrix, which draws upon the best-practice in feature selection while assigning scores according to how well features discriminate between classes. The method (supervised) estimates some indicators of centrality identifying the most important features within the graph. The results are remarkable: the proposed method has been extensively tested on 7 different datasets selected from different scenarios (i.e., object recognition, handwritten recognition, biological data, and synthetic testing datasets), in all the cases we achieve top performances against 7 competitors selected from recent literature in feature selection. Our approach is also robust and stable on different splits of the training data, it performs effectively in ranking high the most relevant features, and it has a very competitive complexity. This study also points to many future directions; focusing on the investigation of different implementations for parallel computing for big data analysis or focusing on the investigation of different relations among the features. Finally, we provide an open and portable library of feature selection algorithms, integrating the methods with uniform input and output formats to facilitate large scale performance evaluation. The *Feature Selection Library* (FSLib is available on *Matlab File Exchange* at https://goo.gl/bvg1ha) and interfaces are fully documented. The library integrates directly with MATLAB, a popular language for machine learning and pattern recognition research.

References

1. GINA digit recognition database. In: IEEE Conference International Joint Conference on Neural Networks (2007)
2. Alon, U., Barkai, N., Notterman, D.A., Gish, K., Ybarra, S., Mack, D., Levine, A.J.: Broad patterns of gene expression revealed by clustering analysis of tumor and normal colon tissues probed by oligonucleotide arrays. PNAS **96**(12), 6745–6750 (1999)
3. Bamber, D.: The area above the ordinal dominance graph and the area below the receiver operating characteristic graph. J. Math. Psychol. **12**(4), 387–415 (1975)
4. Battiti, R.: Using mutual information for selecting features in supervised neural net learning. IEEE Trans. Neural Netw. **5**(4), 537–550 (1994)

5. Bólon-Canedo, V., Sánchez-Maroo, N., Alonso-Betanzos, A.: Recent advances and emerging challenges of feature selection in the context of big data. Knowl.-Based Syst. **86**, 33–45 (2015)

6. Bonacich, P.: Power and centrality: a family of measures. Am. J. Sociol. **92**(5), 1170–1182 (1987)

7. Bradley, P.S., Mangasarian, O.L.: Feature selection via concave minimization and support vector machines. In: Conference International Conference on Machine Learning (ICML) (1998)

8. Duch, W., Wieczorek, T., Biesiada, J., Blachnik, M.: Comparison of feature ranking methods based on information entropy. In: IJCNN, vol. 2. IEEE (2004)

9. Everingham, M., Van Gool, L., Williams, C.K.I., Winn, J., Zisserman, A.: The PASCAL Visual Object Classes Challenge 2007 (VOC2007) Results (2007)

10. Garrison, W.L.: Connectivity of the interstate highway system. Pap. Reg. Sci. **6**(1), 121–137 (1960)

11. Golub, T.R.: Molecular classification of cancer: class discovery and class prediction by gene expression monitoring. Science **286**(5439), 531–537 (1999)

12. Grinblat, G.L., Izetta, J., Granitto, P.M.: SVM based feature selection: why are we using the dual? In: Conference Ibero-American Conference on AI (2010)

13. Gu, Q., Li, Z., Han, J.: Generalized fisher score for feature selection. In: Computing Research Repository (CoRR) (2012)

14. Guyon, I.: Feature Extraction: Foundations and Applications, vol. 207. Springer Science & Business Media, Berlin (2006)

15. Guyon, I., Gunn, S., Ben-Hur, A., Dror, G.: Result analysis of the nips 2003 feature selection challenge. In: NIPS, pp. 545–552 (2004)

16. Guyon, I., Li, J., Mader, T., Pletscher, P.A., Schneider, G., Uhr, M.: Competitive baseline methods set new standards for the NIPS 2003 feature selection benchmark. PRL **28**(12), 1438–1444 (2007)

17. Guyon, I., Weston, J., Barnhill, S., Vapnik, V.: Gene selection for cancer classification using support vector machines. Mach. Learn. J. **46**(1), 389–422 (2002)

18. Guzmán-Martínez, R., Alaiz-Rodríguez, R.: Feature selection stability assessment based on the Jensen-Shannon divergence. In: Gunopulos, D., Hofmann, T., Malerba, D., Vazirgiannis, M. (eds.) ECML PKDD 2011. LNCS, vol. 6911, pp. 597–612. Springer, Heidelberg (2011). doi:10.1007/978-3-642-23780-5_48

19. He, X., Cai, D., Niyogi, P.: Laplacian score for feature selection. In: Advances in Neural Information Processing Systems, vol. 18 (2005)

20. Kang, U., Papadimitriou, S., Sun, J., Tong, H.: Centralities in large networks: algorithms and observations. In: Proceedings of the 2011 SIAM International Conference on Data Mining. Society for Industrial and Applied Mathematics, pp. 119–130 (2011)

21. Kuncheva, L.I.: A stability index for feature selection. In: Proceedings of the 25th Conference on Proceedings of the 25th IASTED International Multi-Conference: Artificial Intelligence and Applications, AIAP 2007, pp. 390–395. ACTA Press, Anaheim (2007)

22. Lehoucq, R.B., Sorensen, D.C., Yang, C.: ARPACK Users' Guide: Solution of Large-Scale Eigenvalue Problems with Implicitly Restarted Arnoldi Methods, vol. 6. SIAM, Philadelphia (1998)

23. Lerman, K., Ghosh, R., Kang, J.H.: Centrality metric for dynamic networks. In: Proceedings of the Eighth Workshop on Mining and Learning with Graphs, MLG 2010, pp. 70–77. ACM, New York (2010)

24. Liu, H., Motoda, H. (eds.): Computational Methods of Feature Selection. CRC Press, Boca Raton (2007)

25. Meyer, C.D. (ed.): Matrix Analysis and Applied Linear Algebra. Society for Industrial and Applied Mathematics, Philadelphia (2000)

26. Obertino, S., Roffo, G., Granziera, C., Menegaz, G.: Infinite feature selection on shore-based biomarkers reveals connectivity modulation after stroke. In: 2016 International Workshop on Pattern Recognition in Neuroimaging (PRNI), pp. 1–4, June 2016

27. Peng, H., Long, F., Ding, C.: Feature selection based on mutual information: criteria of max-dependency, max-relevance, and min-redundancy. IEEE Trans. Pattern Anal. Mach. Intell. (PAMI) **27**(8), 1226–1238 (2005)

28. Pitts, F.R.: A graph theoretic approach to historical geography. Prof. Geogr. **17**(5), 15–20 (1965)

29. Rawat, A., Saha, S., Ghrera, S.P.: Time efficient ranking system on map reduce framework. In: 2015 Third International Conference on Image Information Processing (ICIIP), pp. 496–501 (2015)

30. Roffo, G., Melzi, S.: Online feature selection for visual tracking. In: International Conference the British Machine Vision Conference (BMVC), September 2016

31. Roffo, G., Melzi, S., Cristani, M.: Infinite feature selection. In: IEEE Conference International Conference on Computer Vision (ICCV) (2015)

32. Simonyan, K., Zisserman, A.: Very deep convolutional networks for large-scale image recognition. CoRR abs/1409.1556 (2014)

33. Singh, D., Febbo, P.G., Ross, K., Jackson, D.G., Manola, J., Ladd, C., Tamayo, P., Renshaw, A.A., D'Amico, A.V., Richie, J.P., Lander, E.S., Loda, M., Kantoff, P.W., Golub, T.R., Sellers, W.R.: Gene expression correlates of clinical prostate cancer behavior. Cancer Cell **1**(2), 203–209 (2002)

34. Wu, D.D., Deng, X., Li, Y.: Safety and emergency systems engineering mapreduce based betweenness approximation engineering in large scale graph. Syst. Eng. Procedia **5**, 162–167 (2012)

35. Zaffalon, M., Hutter, M.: Robust feature selection using distributions of mutual information. In: Conference International Conference on Uncertainty in Artificial Intelligence (UAI) (2002)

36. Zhang, Z., Hancock, E.R.: A graph-based approach to feature selection. In: Jiang, X., Ferrer, M., Torsello, A. (eds.) GbRPR 2011. LNCS, vol. 6658, pp. 205–214. Springer, Heidelberg (2011). doi:10.1007/978-3-642-20844-7_21

Classification and Prediction

Bayesian Confirmation Measures in Rule-Based Classification

Dariusz Brzezinski$^{(\boxtimes)}$, Zbigniew Grudziński, and Izabela Szczęch

Institute of Computing Science, Poznan University of Technology,
ul. Piotrowo 2, 60–965 Poznan, Poland
{dariusz.brzezinski,izabela.szczech}@cs.put.poznan.pl

Abstract. With the rapid growth of available data, learning models are also gaining in sizes. As a result, end-users are often faced with classification results that are hard to understand. This problem also involves rule-based classifiers, which usually concentrate on predictive accuracy and produce too many rules for a human expert to interpret. In this paper, we tackle the problem of pruning rule classifiers while retaining their descriptive properties. For this purpose, we analyze the use of confirmation measures as representatives of interestingness measures designed to select rules with desirable descriptive properties. To perform the analysis, we put forward the CM-CAR algorithm, which uses interestingness measures during rule pruning. Experiments involving 20 datasets show that out of 12 analyzed confirmation measures c_1, F, and Z are best for general-purpose rule pruning and sorting. An additional analysis comparing results on balanced/imbalanced and binary/multi-class problems highlights also N, S, and c_3 as measures for sorting rules on binary imbalanced datasets. The obtained results can be used to devise new classifiers that optimize confirmation measures during model training.

Keywords: Rule classifiers · Interestingness measures · Bayesian confirmation · Rule pruning

1 Introduction

Recent years have seen the rise of such terms as big data and data science, which brought many machine learning and data mining methods to public attention. This growing popularity of pattern mining methods results in numerous practical applications, such as healthcare, online education, social network analysis, or smart houses [18,20]. Many of these applications involve cooperation with human experts, who often have to understand not only direct algorithm results, but also entire learning models.

Arguably the most studied data mining task is classification [18]. Various types of classifiers have been developed over the years, however rules are continuously regarded as one of the most popular approaches to practical applications involving non-data-mining experts. It is due to the symbolic form of rules, which

© Springer International Publishing AG 2017
A. Appice et al. (Eds.): NFMCP 2016, LNAI 10312, pp. 39–53, 2017.
DOI: 10.1007/978-3-319-61461-8_3

makes them comprehensible. Thus, when both pattern usage and understanding are key goals, rules are a common form of knowledge representation.

Nevertheless, in most studies data miners tend to focus solely on the predictive performance of learning models [2,6,13]. This is also the case of rule mining. As a result, the descriptive value that rules can carry is often neglected. Unquestionably, a compilation of good predictive and descriptive abilities of a classifier is sought for in many applications. Preferably, these abilities should also be accompanied by a compact representation. In particular, for rule-based classifiers this requirement can be achieved by limiting the number of rules, since otherwise the set of rules could exceed the human-expert's understanding capabilities. For example, in medical applications, doctors are usually interested in a reduced set of rules that describes the patients well and offers good predictions [26].

The evaluation and, thus, pruning of rule sets is usually done by interestingness measures; for a survey see e.g. [14,24]. In classification, these measures are used to improve the predictive performance of learning models, often neglecting the descriptive value of each rule. Nonetheless, many interestingness measures were designed especially for evaluating the descriptive properties of rules. In particular, Bayesian confirmation measures [12] constitute a group of measures that quantify the degree with which the rule's premise supports the conclusion. Confirmation measures obtain positive values only when the premise widens our knowledge about the conclusion, thus, they allow to swiftly choose meaningful rules and filter out the misleading ones. Additionally, the usefulness of confirmation measures in the descriptive context has been depicted with many desirable properties they possess [7,12,15,16].

In this paper, we analyze the impact of using confirmation measures in rule-based classification. For this purpose, we put forward the CM-CAR algorithm, which uses confirmation measures to sort and prune a list of rules. As a result, the proposed algorithm is capable of producing a concise set of descriptive rules, while retaining high predictive performance. Summarizing, the main contributions of this paper are as follows:

- the analysis of interestingness measures with good descriptive properties in the context of predictive classification problems;
- the proposal of the CM-CAR algorithm for discovering and pruning decision rules with high confirmation;
- a comprehensive series of experiments analyzing 12 Bayesian confirmation measures for sorting and pruning rule lists.

The remainder of this paper is organized as follows. Section 2 provides basic notation, definitions, reviews Bayesian confirmation measures, and discusses related works. Section 3 presents the CM-CAR algorithm. In Sect. 4, we discuss experimental results, which demonstrate the properties of the analyzed measures. Finally, Sect. 5 concludes the paper and draws lines of future research.

2   Preliminaries and Related Works

Among various knowledge representations, patterns in the form of rules are known and appreciated for their high comprehensibility and interpretability. Such form of knowledge representation is often found easy to understand and use by decision makers.

Rules are usually induced from a dataset being a set of objects characterized by a set of attributes. Rules are consequence relations, denoted as $E \rightarrow H$ ("*if E then H*"), between the condition E and conclusion H formulas built from attribute-value pairs. The condition formulas are called the premise (or evidence) and the conclusion formulas are referred to as the conclusion (or hypothesis) of the rule. If the set of attributes that can occur in the conclusion is limited to a predefined *class attribute*, then the rule is regarded as a *decision rule*.

The evaluation of the quality and utility of rules induced from data is most commonly done by means of *interestingness measures*, which quantify the relationship between E and H. In the context of a particular dataset, interestingness measures can be usually defined on the basis of four non-negative values: a, b, c and d, briefly represented in Table 1.

Table 1. An exemplary contingency table of the rule's premise and conclusion

	H	$\neg H$	Σ
E	a	c	$a + c$
$\neg E$	b	d	$b + d$
Σ	$a + b$	$c + d$	n

The number of objects in a dataset that satisfy both the rule's premise and conclusion is quantified by a. The number of objects for which the premise is not satisfied, but the conclusion is, will be denoted by b, etc. This notation can be effectively used for defining such interestingness measures as, for example, confidence: $conf(H, E) = a/(a + c)$ or support: $sup(H, E) = a$.

In this paper we focus on a particular group of interestingness measures that are referred to as *Bayesian confirmation measures* (or simply *confirmation measures*). Their common feature is that they obtain:

- positive values when $P(H|E) > P(H)$,
- 0 when $P(H|E) = P(H)$,
- negative values when $P(H|E) < P(H)$.

Observe that the notation based on a, b, c, and d can also be used to estimate probabilities, e.g. $P(H) = (a+b)/n$ or $P(H|E) = a/(a+c)$. Thus, the conditions that a confirmation measure, denoted as $cm(H, E)$, must satisfy can be expressed as follows:

$$cm(H, E) \begin{cases} > 0 \text{ when } \frac{a}{a+c} > \frac{a+b}{n}, \\ = 0 \text{ when } \frac{a}{a+c} = \frac{a+b}{n}, \\ < 0 \text{ when } \frac{a}{a+c} < \frac{a+b}{n}. \end{cases} \tag{1}$$

Thus, confirmation measures quantify the degree to which E provides support *for* or *against* H [12].

Due to the fact that the above conditions do not favor any single measure as the most adequate, there are many alternative, ordinally non-equivalent measures of confirmation [7,12]. Definitions of 12 popular confirmation measures are listed in Table 2.

Table 2. Popular confirmation measures

$D(H,E) = P(H\|E) - P(H) = \dfrac{a}{a+c} - \dfrac{a+b}{n} = \dfrac{ad-bc}{n(a+c)}$	[11]
$M(H,E) = P(E\|H) - P(E) = \dfrac{a}{a+b} - \dfrac{a+c}{n} = \dfrac{ad-bc}{n(a+b)}$	[25]
$S(H,E) = P(H\|E) - P(H\|\neg E) = \dfrac{a}{a+c} - \dfrac{b}{b+d} = \dfrac{ad-bc}{(a+c)(b+d)}$	[5]
$N(H,E) = P(E\|H) - P(E\|\neg H) = \dfrac{a}{a+b} - \dfrac{c}{c+d} = \dfrac{ad-bc}{(a+b)(c+d)}$	[27]
$C(H,E) = P(E \wedge H) - P(E)P(H) = \dfrac{a}{n} - \dfrac{(a+c)(a+b)}{n^2} = \dfrac{ad-bc}{n^2}$	[3]
$F(H,E) = \dfrac{P(E\|H) - P(E\|\neg H)}{P(E\|H) + P(E\|\neg H)} = \dfrac{\dfrac{a}{a+b} - \dfrac{c}{c+d}}{\dfrac{a}{a+b} + \dfrac{c}{c+d}} = \dfrac{ad-bc}{ad+bc+2ac}$	[21]
$Z(H,E) = \begin{cases} 1 - \dfrac{P(\neg H\|E)}{P(\neg H)} = \dfrac{ad-bc}{(a+c)(c+d)} & \text{in case of confirmation} \\[2ex] \dfrac{P(H\|E)}{P(H)} - 1 = \dfrac{ad-bc}{(a+c)(a+b)} & \text{in case of disconfirmation} \end{cases}$	[7]
$A(H,E) = \begin{cases} \dfrac{P(E\|H) - P(E)}{1 - P(E)} = \dfrac{ad-bc}{(a+b)(b+d)} & \text{in case of confirmation} \\[2ex] \dfrac{P(H) - P(H\|\neg E)}{1 - P(H)} = \dfrac{ad-bc}{(b+d)(c+d)} & \text{in case of disconfirmation} \end{cases}$	[16]
$c_1(H,E) = \begin{cases} \alpha + \beta A(H,E) & \text{in case of confirmation when } c = 0 \\ \alpha Z(H,E) & \text{in case of confirmation when } c > 0 \\ \alpha Z(H,E) & \text{in case of disconfirmation when } a > 0 \\ -\alpha + \beta A(H,E) & \text{in case of disconfirmation when } a = 0 \end{cases}$	[16]
$c_2(H,E) = \begin{cases} \alpha + \beta Z(H,E) & \text{in case of confirmation when } b = 0 \\ \alpha A(H,E) & \text{in case of confirmation when } b > 0 \\ \alpha A(H,E) & \text{in case of disconfirmation when } d > 0 \\ -\alpha + \beta Z(H,E) & \text{in case of disconfirmation when } d = 0 \end{cases}$	[16]
$c_3(H,E) = \begin{cases} A(H,E)Z(H,E) & \text{in case of confirmation} \\ -A(H,E)Z(H,E) & \text{in case of disconfirmation} \end{cases}$	[16]
$c_4(H,E) = \begin{cases} min(A(H,E), Z(H,E)) & \text{in case of confirmation} \\ max(A(H,E), Z(H,E)) & \text{in case of disconfirmation} \end{cases}$	[16]

The selected confirmation measures obtain values ranging from -1 to $+1$, except for measures $D(H,E)$ and $M(H,E)$, whose values approach -1 or $+1$

only for n approaching $+\infty$. Moreover, measure $C(H, E)$ originally obtains values from $-1/4$ to $+1/4$ (regardless of n), so a simple linear transformation (a multiplication by 4) has been introduced and all further results concern the transformed $C(H, E)$. For brevity and clarity of presentation, the definitions of measures $Z(H, E)$, $A(H, E)$, $c_1(H, E)$, $c_2(H, E)$, $c_3(H, E)$ and $c_4(H, E)$ in Table 2 omit the situation of neutrality, in which the measures default to 0. Moreover, measures $c_1(H, E)$ and $c_2(H, E)$ have been computed for the values of $\alpha = \beta = 1/2$.

Our interest in confirmation measures results mostly from their valuable scale semantics. Notice, how easy it is to filter out misleading rules (i.e., those for which the premise actually disconfirms the conclusion) only by observing the value of the measure. Especially when working with imbalanced data, it is important not to give credit to rules in which the probability of the conclusion given the premise is smaller than the genuine probability of the conclusion itself. Nevertheless not all popular interestingness measures depict such situations, e.g. confidence, support. That is why, we direct our interest to confirmation measures. They have been widely studied as measures in single-rule evaluation [7,12,16] for descriptive purposes, neglecting however their potential usefulness in classifiers. Thus, our experimental study intentionally focuses only on confirmation measures, which in our opinion should gain in popularity in the context of rule-based classification.

Although classical approaches to rule classification concentrate on predictive performance and rule coverage [6,9,13,28], there have already been studies on using interestingness measures in rule-based classification. The algorithm that particularly inspired our work is CBA [23]. The Classification Based on Associations (CBA) algorithm is based on applying association rule induction approaches to finding classification rules. In CBA the classifier construction process starts by generating association rules characterized by minimal support. Next, the obtained associations are transformed to classification rules by selecting only those rules where the conclusion is the class attribute. Furthermore, these classification rules are filtered and limited only to those with confidence equal or greater than a user-defined threshold. Finally, the set of rules is ordered on the basis of their confidence, support, and length.

Other attempts to use frequent patterns/association rules in classification include the CAEP classifier [10], which is based on emerging patterns. Emerging patterns are defined as patterns whose supports increase significantly from one class to another and, as the CAEP method shows, prove to work well even with high dimensional problems [10]. Among more recent proposals, Ceci and Appice [4] focus on propositional and structural approaches to spatial classification in multi-relational data mining. This work also studies an associative classification framework, one that employs spatial association rules. Nevertheless, none of the cited works investigates the use of Bayesian confirmation measures, which are the main focus of this paper.

3   The CM-CAR Algorithm

In this paper, we analyze the potential of using confirmation measures in classification. However, existing rule classifiers [6,9,13,28] try to optimize accuracy or instance coverage rather than the descriptive value of the created rules. Therefore, we put forward a new algorithm called Confirmation Measure Class Association Rules (CM-CAR), which creates a user-defined number of decision rules based on Bayesian confirmation measures. The pseudocode of CM-CAR is presented in Algorithm 1.

Algorithm 1. CM-CAR

Input: $\mathcal{D}$: data set, *minsup*: minimal support, k: number of rules, C: class attribute, Q_s: ordered set of sorting measures, Q_p: ordered set of pruning measures
Output: $\mathcal{CAR}$: decision rule list of length k
1: $\mathcal{CAR} \leftarrow \emptyset$
2: $\mathcal{L} \leftarrow$ itemsets with support $\geq$ *minsup*   ▷ Find frequent associations
3: **for all** subsets l_k of itemsets $l \in \mathcal{L}$ **do**   ▷ Create decision rules
4:      **if** $l - l_k = \{C\}$ **then**
5:          $r \leftarrow$ decision rule $l_k \rightarrow C$
6:          $\mathcal{CAR} \leftarrow \mathcal{CAR} \cup r$
7: Sort $\mathcal{CAR}$ according to Q_s ▷ Create decision list
8: Leave in $\mathcal{CAR}$ k-best rules according to Q_p ▷ Prune decision list

---

First, the CM-CAR algorithm finds frequent itemsets. For this purpose we use the Apriori algorithm [1], however, in practice any frequent itemset mining algorithm could be used. Next, CM-CAR creates decision rules based on those frequent sets that contain the class attribute C. Finally, two sets of interestingness measures, Q_s and Q_p, are used to sort and filter the rules, respectively. As its classification model, the algorithm outputs a list of k decision rules, where k is a user-defined value.

CM-CAR can be considered a generalization of the CBA algorithm proposed by Liu et al. [23], where instead of using support and confidence, we use arbitrary interestingness measures to create a list of rules. As in the CBA algorithm, the time performance of CM-CAR depends mostly on the frequent pattern mining phase which has a complexity of $O(2^n)$, n being the dataset size.

It is worth noting that the proposed algorithm uses two sets of measures for two distinct purposes. Q_s is a set of measures that order the rules and, therefore, decide which rule is used if more than one rule covers an example. If $Q_s = \{sup, conf\}$, rules are sorted according to their support and then, in case of ties, confidence. On the other hand, Q_p prunes the sorted rules. For example, if $Q_p = \{S, N\}$ then the rule list is limited to k best rules according to measure S and, in case of ties, N.

With two separate sets of measures, CM-CAR is capable of dividing the responsibility for the predictive (Q_s) and descriptive (Q_p) properties of its classification model. In the following section, we use this feature to compare various confirmation measures in a series of experiments.

4   Experimental Study

The goal of this paper is to perform a comparison of confirmation measures. For this purpose, we use the CM-CAR algorithm with varying values of Q_s and Q_p. The use of other rule-based classifiers is out of the scope of this study.

The experiments are divided into two groups. In the first group, we are interested in assessing confirmation measures in the context of rule pruning. Therefore, we set $Q_s = \{conf, sup, length\}$ and $Q_p = \{CM\}$, where *length* signifies the number of conditional attributes in a rule and CM is one of the 12 confirmation measures from Table 2. For reference, we also analyzed the usage of *conf* as a pruning measure. It is worth noting that using *conf* for pruning makes CM-CAR work exactly like the CBA algorithm. Therefore, *conf* can be considered a baseline against which the remaining measures can be compared. By keeping Q_s fixed in this group of experiments, we ensure that differences in model performance are only due to the measure used for pruning.

In the second group of experiments, we focus on verifying the utility of confirmation measures in the context of classification. To achieve this, we set $Q_s = \{CM, sup, length\}$ and $Q_p = \{CM\}$, making one of the 12 confirmation measures (or *conf*) a key factor responsible for the predictive and descriptive performance. As in the first group of experiments, *conf* serves as a baseline approach against which other measures can be compared.

The *minsup* parameter for frequent pattern mining was set to obtain a number of rules close to 10 000 for each dataset. Such a number was selected to ensure that it is possible to perform a long series of rule prunings. The use of each confirmation measure was evaluated on a holdout test set consisting of 34% of the original dataset using [19]:

- Balanced accuracy: $\frac{1}{2}$(sensitivity + specificity),
- G-mean: $\sqrt{\text{sensitivity} \cdot \text{specificity}}$,
- F_1-score: $2 \cdot \frac{\text{sensitivity} \cdot \text{precision}}{\text{precision} + \text{sensitivity}}$,
- AUC: area under the Receiver Operator Characteristic curve [19].

For multi-class problems, performance was calculated using macro averaging, i.e., evaluation measures where computed "one-vs-all" for each class and averaged without weighting. All four measures were chosen based on their ability to assess classifiers on imbalanced data. The CM-CAR algorithm was written in Java as part of the WEKA [17] framework.[1]

4.1   Datasets

In our study, we used 20 datasets with various numbers of classes, imbalance ratios, and containing nominal as well as numeric attributes. All of the used datasets are publicly available, mostly through the UCI machine learning repository [22]. Table 3 presents the main characteristics of each dataset.

---

[1] Sources available at: http://www.cs.put.poznan.pl/dbrzezinski/software.php.

Table 3. Dataset characteristics

Dataset	Size	Num. attr.	Nom. attr.	Classes	Maj. class	Mined rules	Balanced	Binary
adult-census	32,561	6	8	2	75.90%	4,299	✗	✓
autos	205	15	10	7	32.68%	8,109	✗	✗
cmc	1,473	2	7	3	43.70%	10,001	✓	✗
credit-g	1,000	7	13	2	70.00%	8,540	✗	✓
diabetes	768	8	0	2	64.10%	10,085	✓	✓
electricity	45,312	7	1	2	57.50%	9,210	✓	✓
hepato	536	9	0	4	33.20%	5,055	✓	✗
king-and-rook	28,056	0	6	18	16.20%	10,266	✗	✗
kr-vs-kp	3,196	0	36	2	52.20%	10,542	✓	✓
lymph	148	3	15	4	54.73%	8,934	✗	✗
madelon	2,600	500	0	2	50.00%	2,431	✓	✓
mushroom	8,124	0	22	2	51.80%	6,468	✓	✓
nursery	12,960	0	8	5	33.30%	9,642	✗	✗
poker-hand	829,201	5	5	10	50.10%	9,267	✗	✗
spect	267	0	22	2	79.40%	9,290	✗	✓
splice	3,190	0	61	3	51.88%	8,313	✗	✗
tic-tac-toe	958	0	9	2	65.34%	9,134	✗	✓
vowel	990	10	3	11	9.09%	6,921	✓	✗
waveform	5,000	40	0	3	33.80%	10,644	✓	✗
wine	153	13	0	3	39.87%	4,697	✓	✗

Out of all the datasets, 10 can be considered balanced, whereas 10 suffer from class-imbalance. Similarly, 9 datasets represent binary classification problems, while 11 have more than two classes. Most datasets have from few hundred to few thousand examples, with the notable exception of poker-hand which contains 829,201 instances. It is also worth highlighting madelon as the dataset with most descriptive attributes (500) and king-and-rook as the one with most class attribute values (18).

Due to the fact that CM-CAR creates rules from frequent itemsets, it requires instances described only by nominal attributes. Therefore, all numerical attributes were discretized into ten equal-frequency bins. Datasets preprocessed in this way were used in all the discussed experiments.

4.2 Rule Pruning

In this group of experiments, the generated rule set was pruned subsequently by: 0%, 10%, 20%, 30%, 40%, 50%, 60%, 70%, 80%, 90%, 95%, 98%, 99% of the original model size. Thus, at the extremes the rule set was not pruned at all or was limited to only 1% of the initial set. Due to the large number of tested measures and datasets, we will only present the most interesting results; detailed tables and additional plots are available in the supplementary materials.[2]

---

[2] Supplement: http://www.cs.put.poznan.pl/dbrzezinski/software/CMCAR.html.

For evaluations using the G-mean measure, it was observed that since G-mean multiplies the true positive rate of each class, in situations where the rules did not cover examples from one of the classes the reported performance was zero. This shows that for highly imbalanced data coverage should be additionally controlled. Partially due to this phenomenon, on some of the datasets (madelon, spect, tic-tac-toe, poker-hand, kr-vs-kp, king-and-rook) the differences in performance were very small and did not discriminate confirmation measures in terms of pruning capabilities. However, on the remaining data clear differences were visible, and two cohesive groups of measures were identified: (1) A and c_2; (2) F, Z, and c_1. Figure 1 presents measure performance on two datasets, which exemplify the relations between these two groups.

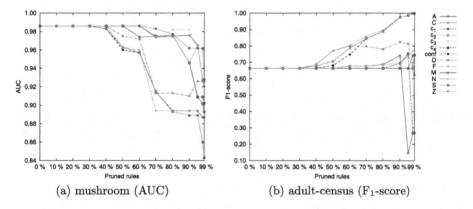

(a) mushroom (AUC)   (b) adult-census (F$_1$-score)

Fig. 1. CM-CAR's AUC on the mushroom dataset and F$_1$-score on adult-census for different pruning levels with $Q_s = \{conf, sup, length\}$ and $Q_p = \{CM\}$, where CM is one of the measures listed in the legend.

The dependency between measures A and c_2 can be explained by the fact that the value of c_2 is in some cases proportional to the value of A. Such a situation occurs in the case of confirmation and when additionally b (the number of objects not supporting the premise, but supporting the conclusion) is greater than 0. Indeed, analyzing the obtained frequent itemsets we noticed that these two requirements were met for most datasets.

The relation between measures in the second group is more difficult to explain. Under certain conditions, c_1 is proportional to Z, however the interdependence with F is not expressed in any way in the definitions of these measure. It is worth noting that all three measures were among the best performing pruning measures, when evaluated using balanced accuracy, G-mean, AUC, and F$_1$-score.

To verify the significance of the observed differences, we performed the nonparametric Friedman test [8]. The null-hypothesis of the Friedman test (that there is no difference between the performance of all the tested confirmation measures) can be rejected for balanced accuracy, G-mean, and the F$_1$-score with $p < 0.05$, but not for AUC. To verify which confirmation measures perform better than the other, we computed the critical difference (CD) chosen by the Nemenyi post-hoc test [8] at $\alpha = 0.05$. Figure 2 depicts the results of the test for balanced accuracy, F_1-score, and G-mean by connecting the groups of measures that are not significantly different (the lower the rank the better).

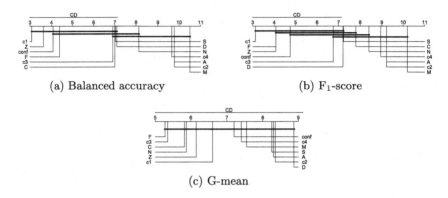

(a) Balanced accuracy   (b) F_1-score

(c) G-mean

Fig. 2. Performance ranking of all measures ($Q_s = \{conf, sup, length\}$, $Q_p = \{CM\}$) averaged over all the analyzed pruning levels. Measures that are not significantly different according to the Nemenyi test (at $\alpha = 0.05$) are connected.

As mentioned earlier, F, Z, c_1 are among the best measures according to balanced accuracy and the F_1-score. Similar rankings were found for G-mean, however, due to the large number of compared measures the post-hoc test for these measures was unable to distinguish groups of measures performing significantly differently. For balanced accuracy and F_1-score, the test was not able to showcase a significant difference with $conf$, S, D and c_3, however, at $\alpha = 0.05$ the three highlighted measures pruned significantly better than C, N, c_4, M, c_2, and A. It is also worth noticing, that according to G-mean $conf$ performs much worse than according to balanced accuracy or F_1-score. This may suggest that $conf$ promotes focusing on overall accuracy potentially neglecting underrepresented minority classes.

4.3   Classification Using Confirmation Measures

In the second group of experiments, we used confirmation measures to sort the rule list and, thus, influence the classification procedure. Tables with balanced accuracy, G-mean, AUC, and F_1-score performance for CM-CAR using each of the analyzed measures are available in the supplementary material (See footnote 2), whereas below we summarize the main findings.

In terms of average predictive performance for all pruning levels, F, Z, c_1 were once again the best performing measures. It is also worth highlighting S and c_3, which were also among the best measures. This is particularly interesting as these measures possess desirable properties, such as minimality/maximality or evidence symmetry and evidence-hypothesis symmetry, which are not showcased by F, Z, or c_1 [16]. Another consistent observation was that of M, A, and c_2 being the worst measures for rule sorting. Two exemplary datasets where these relations can be seen are presented in Fig. 3.

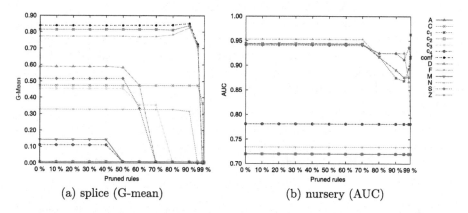

(a) splice (G-mean)   (b) nursery (AUC)

Fig. 3. CM-CAR's G-mean on the splice dataset and AUC on nursery for different pruning levels with $Q_s = \{CM, sup, length\}$ and $Q_p = \{CM\}$.

As in the first group of experiments, we performed the Friedman test. The null-hypothesis of the Friedman test can be rejected for all four evaluation measures (balanced accuracy, G-mean, AUC, F-score) with $p < 0.001$. Figure 4 visually presents the results of the Nemenyi test.

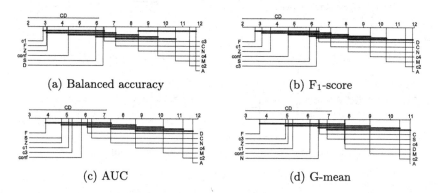

(a) Balanced accuracy   (b) F_1-score

(c) AUC   (d) G-mean

Fig. 4. Performance ranking of all measures ($Q_s = \{CM, sup, length\}$, $Q_p = \{CM\}$) averaged over all the analyzed pruning levels. Measures that are not significantly different according to the Nemenyi test (at $\alpha = 0.05$) are connected.

As the results show, F, Z, c_1 are once again the best measures, and are significantly better at rule sorting than c_4, M, c_2, and A.

4.4   The Impact of Imbalance Data and Multiple Classes

The previous two subsections analyzed the potential of using Bayesian confirmation measures for rule list pruning and sorting. However, datasets selected for this study allow us to differentiate the performance of the measures on balanced/imbalanced and binary/multi-class problems. The last two columns of Table 3 distinguish both types of dataset categorizations.

Figures 5 and 6 present the results of Nemenyi post-hoc tests at $\alpha = 0.05$, with performance on balanced/binary in the left column and imbalanced/multi-class data in the right column. Due to space limitations we only show results for strategies where the confirmation measure was used for both pruning and sorting; for additional plots please refer to the supplementary materials (See footnote 2).

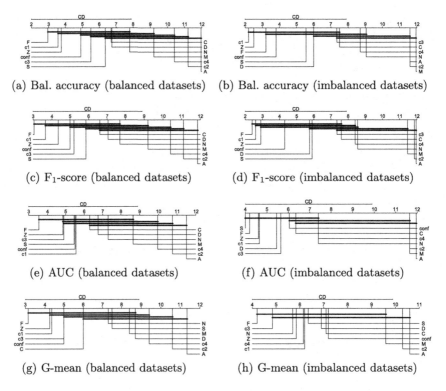

(a) Bal. accuracy (balanced datasets)   (b) Bal. accuracy (imbalanced datasets)

(c) F_1-score (balanced datasets)   (d) F_1-score (imbalanced datasets)

(e) AUC (balanced datasets)   (f) AUC (imbalanced datasets)

(g) G-mean (balanced datasets)   (h) G-mean (imbalanced datasets)

Fig. 5. Performance ranking of all measures ($Q_s = \{CM, sup, length\}$, $Q_p = \{CM\}$) analyzed separately for balanced and imbalanced datasets. Measures that are not significantly different according to the Nemenyi test (at $\alpha = 0.05$) are connected.

Considering balanced datasets, the results are fairly similar to those obtained when analyzing all datasets and highlight F, Z, and c_1. However, when looking at critical distance plots for AUC and G-mean it is also worth mentioning S, N,

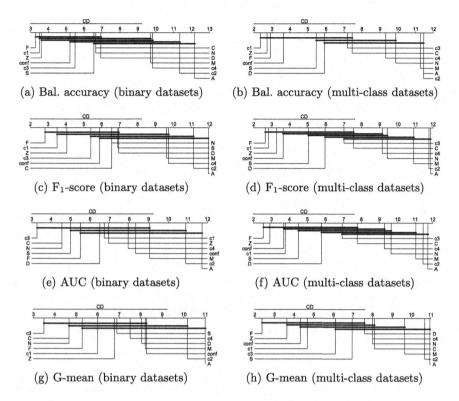

Fig. 6. Performance ranking of all measures ($Q_s = \{CM, sup, length\}$, $Q_p = \{CM\}$) analyzed separately for binary and multi-class problems. Measures that are not significantly different at $\alpha = 0.05$ are connected.

and c_3 as highly ranked measures. This is interesting as all three measures possess minimality/maximality, evidence symmetry, and evidence-hypothesis symmetry properties, mentioned previously [16].

Comparing measure rankings on binary and multi-class problems we can see that most evaluations still promote F, Z, and c_1. A slight deviation from this pattern can be seen on critical distance plots of AUC and G-mean for binary datasets, where c_3, N, and C are the three highest ranked confirmation measures.

5   Conclusions

Mining a concise set of descriptive rules that is characterized by good predictive performance is a challenging task. In this paper, to tackle this problem we proposed the CM-CAR algorithm, which uses confirmation measures to sort and prune a list of rules. Using the proposed algorithm we reviewed the applicability of 12 confirmation measures to rule pruning and sorting.

The results of the experiments show that Bayesian confirmation measures can be successfully applied to reduce the set of rules while maintaining satisfactory predictive performance. In particular, the F, Z, c_1 measures consistently showed better performance than the popularly used *confidence* measure.

An additional analysis comparing results for balanced and imbalanced datasets highlighted N, c_3, and S as promising measures for imbalanced data. This result is particularly interesting as all three measures are well established in the field of interestingness measures and possess additional properties compared to F, Z, c_1, such as: evidence symmetry, evidence-hypothesis symmetry, or minimality/maximality. A similar analysis comparing results for binary and multi-class problems revealed that F, Z, c_1 are ranked highest on both types of problems, with the exception of AUC and G-mean results for binary datasets where c_3, N, and C were the three best confirmation measures.

The results of the research described in this paper inspire us to continue working with confirmation measures in the context of rule-based classification. In particular, we plan to analyze the impact that confirmation measures can have on the coverage of the training set of objects, as in certain applications it is advisable to propose a set of rules that covers the whole or the vast part of the training set. Moreover, based on the results of the comparison performed in this paper, we plan to use selected measures as components of more specialized rule-based classifiers. Finally, a possible extension of CM-CAR can include optimizing the set of classification association rules to those that are not contained by other discovered rules.

Acknowledgements. This work was supported by the National Science Centre grant DEC-2013/11/B/ST6/00963. D. Brzezinski acknowledges the support of an FNP START scholarship and Institute of Computing Science Statutory Fund.

References

1. Agrawal, R., Srikant, R.: Fast algorithms for mining association rules in large databases. In: Proceedings of the 20th International Conference on Very Large Data Bases, pp. 487–499 (1994)
2. Brzezinski, D., Stefanowski, J.: Combining block-based and online methods in learning ensembles from concept drifting data streams. Inf. Sci. **265**, 50–67 (2014)
3. Carnap, R.: Logical Foundations of Probability. University of Chicago Press, Chicago (1962)
4. Ceci, M., Appice, A.: Spatial associative classification: propositional vs structural approach. J. Intell. Inf. Syst. **27**(3), 191–213 (2006)
5. Christensen, D.: Measuring confirmation. J. Philos. **96**, 437–461 (1999)
6. Cohen, W.W.: Fast effective rule induction. In: Proceedings of the 12th International Conference on Machine Learning (ICML 1995), pp. 115–123 (1995)
7. Crupi, V., Tentori, K., Gonzalez, M.: On Bayesian measures of evidential support: theoretical and empirical issues. Philos. Sci. **74**, 229–252 (2007)
8. Demsar, J.: Statistical comparisons of classifiers over multiple data sets. J. Mach. Learn. Res. **7**, 1–30 (2006)
9. Domingos, P.: The rough set based rule induction technique for classification problems. In. In Proceedings of the Sixth IEEE International Conference on Tools with Artificial Intelligence, pp. 704–707 (1994)

10. Dong, G., Zhang, X., Wong, L., Li, J.: CAEP: classification by aggregating emerging patterns. In: Arikawa, S., Furukawa, K. (eds.) DS 1999. LNCS (LNAI), vol. 1721, pp. 30–42. Springer, Heidelberg (1999). doi:10.1007/3-540-46846-3_4

11. Eells, E.: Rational Decision and Causality. Cambridge University Press, Cambridge (1982)

12. Fitelson, B.: The plurality of Bayesian measures of confirmation and the problem of measure sensitivity. Philos. Sci. **66**, 362–378 (1999)

13. Frank, E., Witten, I.H.: Generating accurate rule sets without global optimization. In: Shavlik, J. (ed.) Fifteenth International Conference on Machine Learning, pp. 144–151. Morgan Kaufmann, Burlington (1998)

14. Geng, L., Hamilton, H.: Interestingness measures for data mining: a survey. ACM Comput. Surv. **38**(3) (2006). Article no. 9

15. Glass, D.H.: Confirmation measures of association rule interestingness. Knowl. Based Syst. **44**, 65–77 (2013)

16. Greco, S., Słowiński, R., Szczęch, I.: Properties of rule interestingness measures and alternative approaches to normalization of measures. Inf. Sci. **216**, 1–16 (2012)

17. Hall, M., Frank, E., Holmes, G., Pfahringer, B., Reutemann, P., Witten, I.H.: The weka data mining software: an update. SIGKDD Explor. Newsl. **11**(1), 10–18 (2009)

18. Han, J., Kamber, M., Pei, J.: Data Mining: Concepts and Techniques, 3rd edn. Morgan Kaufmann Publishers Inc., Burlington (2011)

19. Japkowicz, N.: Assessment metrics for imbalanced learning. In: He, H., Ma, Y. (eds.) Imbalanced Learning: Foundations, Algorithms, and Applications, pp. 187–206. Wiley-IEEE Press, Hoboken (2013)

20. Kantardzic, M.: Data-mining applications. In: Data Mining: Concepts, Models, Methods, and Algorithms, 2 edn, pp. 496–509. Wiley (2011)

21. Kemeny, J., Oppenheim, P.: Degrees of factual support. Philos. Sci. **19**, 307–324 (1952)

22. Lichman, M.: UCI machine learning repository (2013). http://archive.ics.uci.edu/ml

23. Liu, B., Hsu, W., Ma, Y.: Integrating classification and association rule mining. In: Proceedings of the 4th International Conference on Knowledge Discovery and Data Mining, pp. 80–86 (1998)

24. McGarry, K.: A survey of interestingness measures for knowledge discovery. Knowl. Eng. Rev. **20**(1), 39–61 (2005)

25. Mortimer, H.: The Logic of Induction. Prentice Hall, Paramus (1988)

26. Napierala, K., Stefanowski, J.: Addressing imbalanced data with argument based rule learning. Expert Syst. Appl. **42**(24), 9468–9481 (2015)

27. Nozick, R.: Philosophical Explanations. Clarendon Press, Oxford (1981)

28. Stefanowski, J.: The rough set based rule induction technique for classification problems. In. In Proceedings of 6th European Conference on Intelligent Techniques and Soft Computing EUFIT, vol. 98 (1998)

Audio-Based Speed Change Classification for Vehicles

Elżbieta Kubera[1], Alicja Wieczorkowska[2(✉)], Tomasz Słowik[3],
Andrzej Kuranc[3], and Krzysztof Skrzypiec[4]

[1] Department of Applied Mathematics and Computer Science,
University of Life Sciences in Lublin, Akademicka 13, 20-950 Lublin, Poland
elzbieta.kubera@up.lublin.pl
[2] Polish-Japanese Academy of Information Technology,
Koszykowa 86, 02-008 Warsaw, Poland
alicja@poljap.edu.pl
[3] Department of Energetics and Transportation,
University of Life Sciences in Lublin, Akademicka 13, 20-950 Lublin, Poland
{tomasz.slowik,andrzej.kuranc}@up.lublin.pl
[4] Maria Curie-Skłodowska University in Lublin,
Pl. Marii Curie-Skłodowskiej 5, 20-031 Lublin, Poland
krzysztof.skrzypiec@poczta.umcs.lublin.pl

Abstract. Vehicle speed is an important factor influencing highway traffic safety. Radars are applied to control the speed of vehicles, but the drivers often decelerate when approaching radar, and then accelerate after passing by. We address automatic recognition of speed change from audio data, based on recordings taken in controlled conditions. Data description and classification experiments illustrate both changing speed and maintaining constant speed. This is a starting point to investigate what percentage of drivers actually maintain constant speed, or slow down only to speed up immediately afterwards. Automatic classification and building an appropriate database can help improving traffic safety.

Keywords: Intelligent transport system · Road traffic safety · Audio signal analysis

1 Introduction

Development of transport brings numerous benefits, but it also brings problems, including pollution of the environment, and decreased safety. More than 1 million people die each year as a result of road traffic crashes, and people aged 15–44 years account for almost a half of global traffic deaths [20]. Pedestrians, cyclists and motorcyclists are especially vulnerable road users. Road crashes also cause economic losses. Additionally, people from low- and middle-income backgrounds and countries are more often involved in road crashes.

Improving the safety of roads and enhancing the behavior of road users became priority in actions related to transport in Poland and around the world.

© Springer International Publishing AG 2017
A. Appice et al. (Eds.): NFMCP 2016, LNAI 10312, pp. 54–68, 2017.
DOI: 10.1007/978-3-319-61461-8_4

Crashes and dangerous situations are analyzed, and the society is informed about the results, which sometimes evoke heated discussions. The use of photo radars, traffic calming zones, speed bumps and other means of improving traffic safety are often criticized by the drivers. However, the statistics confirm that road safety improves, and the number of fatalities in road crashes decreases (Fig. 1). The main factors decreasing safety are:

- low quality of many roads, and of vehicles;
- lack of protection for pedestrians and bikers,
- deficiencies of the road safety systems - prolonged implementation of laws relating to risk factors, insufficient financial support, low public awareness;
- reckless behavior of drivers and other road users - speeding, driving and walking while impaired; not using seat belts.

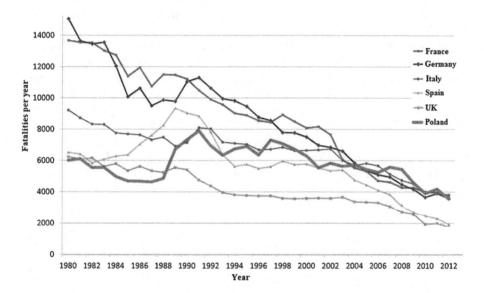

Fig. 1. Fatalities in road traffic crashes in selected European countries [8]

The road safety can be improved through setting and enforcing appropriate regulations, through public awareness campaigns, and interventions targeting the road users behavior. The reports from the operators of navigation systems show that drivers increase speed when passing a speed camera, after slowing down when approaching it. However, these reports describe the behavior of the users of these navigation systems, and further research on other drivers' behavior is needed. The data on drivers' behavior can be collected from audio recording systems, if such systems are developed and deployed. Therefore, the goal of our research is to investigate if changing (or maintaining) speed can be recognized from audio data, as this can be the first step to prepare a system to collect data on drivers' behavior around speed cameras. Vehicle speed also influences the noise level generated by the vehicle, as the speed increase increases the noise level. Other factors increasing the noise level are:

– road inclination - traveling uphill generates more noise, but average traffic speed decreases; traveling downhill generates less noise than on flat road,
– road surface, and its dampness,
– the technical condition of vehicles; especially old vehicles generate more noise and exhaust fumes, and these vehicles also decrease the traffic road safety.

Out of all the factors influencing the road traffic safety and noise generated, the speed of vehicles can be relatively easily addressed, so we decided to address the behavior of drivers when speed is controlled, and prepare audio data representing acceleration, deceleration, and constant speed. These data are very complex, since audio signal depends on many factors, and changes quickly over time. The audio data represent amplitude changes of the audio waveform with time; close-up of a segment of such data is shown in Fig. 2. We can see some periodicity here, but irregular variations are also clearly visible. Also, if the phases of some components of a complex sound change, the wave shape may change dramatically, even though the sound still sounds the same. Therefore, sounds representative for particular sound classes (e.g. for speed increase) do not follow a particular pattern nor trend in time domain. Thus, time domain data are usually parameterized, based on frequency content (i.e. on the spectrum), to capture sound characteristics. The spectrum is usually calculated for short segments of data, to show frequency contents of the sound, and changes of the spectrum over time can be observed in spectrogram, where the amplitude of particular frequencies is represented using a selected color scale, see Fig. 3 (in grayscale).

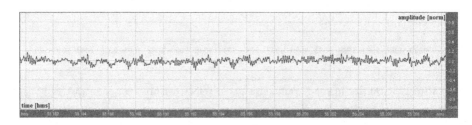

Fig. 2. Audio data in time domain: horizontal axis represents time, vertical - amplitude. Recording for Fiat Ducato (plot from Adobe Audition [1])

Since the vehicle sound contains noise and harmonic components, we designed a feature vector that allows capturing noise features, harmonic features, and their changes in time. Speed is usually monitored by radars, but with a single measurement, and the drivers are aware of this fact. We hope that monitoring how drivers change speed in these circumstances and launching public awareness campaign can increase the road traffic safety.

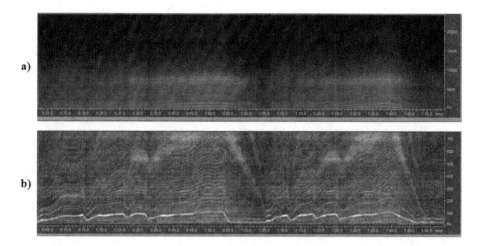

Fig. 3. Spectrogram of Fiat Ducato accelerating, then decelerating, accelerating again and decelerating, for 450 N load (a). A close-up for low frequencies is shown in (b). Higher luminance represents higher amplitude. Plots from Adobe Audition [1]

Audio data has been used in research on automotive control, e.g. for vehicle identification, or fault diagnosis [2, 6, 7, 17], but, to the best of our knowledge, there is no prior research on automatic speed change recognition.

Although audio data represent changes over time, direct time-domain representation is rarely applied in classification. The data represent very fast changes, i.e. thousands of samples per second, and the values of a small number of audio samples are not of particular interest, contrary to typical time series data, where one observation per day or hour or minute is taken. Spectrum or spectrogram are much more useful in audio data analysis [9], but they are still very complex and are usually parameterized before further processing [21]. However, the finding an appropriate parametrization that can capture the features of the target classes of the investigated automotive audio is not an easy task. There is no standard set of parameters that can be used successfully for all research goals in audio data. Also acquiring the automotive data can be challenging, as many factors influence the data values, and the acquired data should be representative for any new data that can be anticipated in tests.

The main contribution of this paper is the parametrization of the vehicle audio data in order to capture the speed changes, or to recognize the speed as approximately stable. Additionally, we performed experiments on time-domain data directly, using time series methods. Also, we prepared the data in controlled conditions. We first recorded vehicles on a dyno test bench, with a constant mic position with respect to the vehicle. Three types of classifiers were carefully tuned for further work. After classification experiments on these data, the on-road recordings were prepared. These data are different than dyno recordings, as the vehicles move with respect the position of the mic, and Doppler effect is present. We describe experiments on dyno data, and then trials on the on-road data, using also dyno data in training, with Doppler effect added. Time domain based and feature set based approaches are compared on a subset of these data, using features designed for this experiment. This illustrates how difficult the task is, and that goal-oriented features can improve the results dramatically.

2   Data

The recordings in controlled conditions at the dyno test bench, with OBD (On-Board Diagnostics) acquisition, were done in May and June 2016, at the University of Life Sciences in Lublin, see Fig. 4. We recorded 8 vehicles:

- Smart ForFour - car with gasoline engine,
- Ford Focus - car with gasoline and LPG (Liquid Petroleum Gas) engine,
- Hyundai i30 - car with Diesel engine,
- Toyota Corolla Verso - car with Diesel engine,
- Daewoo Lublin - van with Diesel engine,
- Fiat Ducato - van with Diesel engine,
- Volkswagen (VW) Transporters, 2004 and 2007 year - vans, Diesel engines.

Fig. 4. Dyno recording. OBD acquisition shown, and the audio recorder (on the tripod)

The vehicles, all with manual transmission, accelerated to 110 km/h (with the exception of Daewoo - to 90 only), then decelerated to low speed at fifth gear, to about 40–45 km/h, when the gear was changed. When accelerating, the driver changed gear and attempted to maintain a constant speed for a few seconds at 50, 70, 90 and 110 km/h. We recorded data (48 kHz/24 bit, stereo) for 3 classes: accelerating, decelerating, and maintaining constant speed. Two versions of dyno loads were recorded: 450 Newton, for all cars with the exception of Ford, and load adjusted to on-road conditions (with the exception of Daewoo and Hyundai), i.e. depending on vehicle speed, weight, and road coefficients.

2.1 On-Road Data

We also prepared on-road recordings in controlled conditions. The mic was positioned as close to the road as possible (Fig. 5), 1.5 m above the road. The recordings were taken on August 2nd 2016, on a sunny summer day (weekday), on a little road in Ciecierzyn, in Lublin Voivodeship in Poland. We chose a little and unfrequented road, to avoid uncontrolled traffic. Unfortunately, the road is not flat here, but the mic was positioned in a mild basin (so the vehicles neither travel uphill nor downhill when passing the mic), with very gentle slopes, similar both to the East and to the West. Three cars were recorded:

– Renault Espace - car with gasoline and LPG engine,
– Toyota Corolla Verso - car with Diesel engine,
– Skoda Octavia - car with Diesel engine.

Fig. 5. The road and microphone position in on-road recordings

The recordings were made on a 100 m long portion of the road. Additionally, video was recorded inside each vehicle, showing the speedometer, in order to facilitate ground-truth labeling of these data. Each vehicle was recorded as follows (twice, eastbound and westbound):

– increasing speed from 50 km/h to 70 km/h,
– decreasing speed from 70 km/h to 50 km/h,
– maintaining stable speed, 50 km/h; additionally, 70 km/h for Skoda.

3 Feature Set

There exist many parameters that can be extracted as audio data features, see [11, 12] for a broad review; these parameters mainly come from speech recognition and musical instrument classification tasks. Still, there is no standard all-purpose feature set, and the feature vector is usually designed for the task in hand. The classification results depend on the feature set used. Since we wanted to capture changes of audio signal properties with speed changes, if any, we decided to observe how sound properties change within 1-s audio frames. Our feature vector consists of 5 groups of features. We are aware that the feature vector is long, but we assumed feature selection when calculating the features. Next, we performed feature selection in these groups.

Basic Features. For each 1 s frame, we calculate spectral features for the starting 330 ms sub-frame, and for the ending 330 ms sub-frame. Next, the calculated features for the starting part, together with the vector of differences between features for the starting and ending part, are placed in the feature vector. Additionally, one time-domain parameter is calculated, i.e. zero-crossing rate, together with its change between the value for the beginning and the ending part. Also, a parameter capturing the spectrum change between the starting and ending part is added to the feature set.

Features describing spectrum and time domain for the starting 330 ms of a •1-s frame include:

- *Zero Crossing Rate* (ZCR) in the time-domain of the sound; a zero-crossing is a point where the sign of the function (amplitude vs. time) changes;
- *Audio Spectrum Envelope* (SE) $SE_0, \ldots SE_{32}$ calculated as sums of the power spectrum coefficients within logarithmically spaced frequency bands [12];
- *SUM_SE* - sum of the spectrum envelope values;
- *MAX_SE_V, MAX_SE_IND* - value and index of the SE maximum;
- *Audio Spectrum Flatness, $flat_1, \ldots, flat_{25}$* - a vector describing the flatness property of the power spectrum [12], i.e. the deviation of the signal's power spectrum from a flat shape, for each band of the spectrum envelope; bands up to SE_{25} were used, as in the research on audio data for vehicles the spectrum is usually limited, so even the sampling rate is often decreased,
- *MFCC* - 13 mel frequency cepstral coefficients (MFCC) [13]. The cepstrum is calculated as the logarithm of the magnitude of the spectral coefficients, and then transformed to the mel scale, reflecting the properties of the human perception of frequency; 24 mel filters were applied, and the obtained results were transformed (averaged) to 12 coefficients. The 13th coefficient is the 0-order coefficient of MFCC, corresponding to the logarithm of the energy;
- *F0_ACor* - fundamental frequency (from the autocorrelation function); this frequency changes with the speed change;
- *Energy* - energy of the entire spectrum;
- *EnAb4kHz* - proportion of the spectral energy above 4 kHz to *Energy*;
- *Audio Spectrum Centroid* (SC) - the power weighted average of the frequency bins in the power spectrum. Coefficients were scaled to an octave scale anchored at 1 kHz [12];
- *Audio Spectrum Spread* (SS) - RMS (root mean square) of the deviation of the log frequency power spectrum wrt. *Audio Spectrum Centroid* [12];
- *RollOff* - the frequency below which 85% (experimentally chosen threshold) of the accumulated magnitudes of the spectrum is concentrated,
- *BW_10dB, BW_20dB, BW_30dB* - bandwidth of the frequency band comprising the spectrum maximum (in dB scale) and the level drop by 10, 20 and 30 dB, respectively, towards both lower and upper frequencies.

All these features are placed in the feature vector. Next, they are also calculated for the ending 330 ms of the 1-s frame, and

- the differences between these values and the values for the starting part are added to the feature vector;

– *Flux1* parameter is added to the feature set. *Flux1* is the sum of squared differences between the magnitudes of the spectrum points calculated for the starting and ending 330 ms sub-frames within the one-second frame.

Altogether, this group of features consists of 169 features (*Basic* group). Fast Fourier transform (FFT) was used for spectrum calculation, and sliding frame with 330 ms hop size was applied when analyzing the audio data and calculating the feature vector. The 330 ms analyzing frame yields sufficient frequency resolution, needed when calculating low frequencies.

One Second Analysis-Based Binned Spectral Features. We decided to add groups of features, representing binned FFT coefficients. This follows [17], with non-stationary automotive data: car engine recordings using mobile phone moved above the engine. They applied FFT on 2.5 s segments, and created a 1000-element feature vector representing averaged magnitudes within 10 Hz bins up to 10 kHz. Since our audio data change too fast to apply 2.5 s analysis, we performed 1 s FFT instead, thus obtaining *FT1s* group (1000 features).

330 ms Analysis-Based Binned Spectral Features. Similarly, we calculated binned spectral features based on 330 ms segments, as this is the frame length used for calculating *Basic* features. Again, 1000 features were calculated (*FT330 ms* group), representing averaged FFT magnitudes within 10 Hz bins up to 10 kHz. The first 330 ms of a 1 s frame was used for calculating these features.

330 ms Analysis-Based Spectral Differences. As our goal is to capture changes over time, we also calculated features describing differences between the starting and ending 330 ms of each analyzed 1 s frame. To this end, we additionally calculated 330 ms binned spectral features for the ending 330 ms. Next, we calculated differences between the corresponding bins of the starting and ending 330 ms frames, yielding 1000 features (*dFT330 ms* group).

3.1   Feature Selection

Since our feature vector is large, we also performed feature selection, as it is recommended in such a case [10]. In our first experiments (on dyno data), we performed feature selection and classification for the basic features only (169 features). For each of the classifiers investigated, cross validation (CV) procedure was applied. We applied 8-fold CV, and each fold represented data from one vehicle. Since we used random forests, we decided to apply feature importance from these classifiers. We tested 2 versions: with constant number of features to be selected (10 features; number arbitrarily chosen), and with feature importance above various thresholds based on mean decrease of Gini criterion. The threshold 0.01 was selected, as giving a small number of features. After threshold based selection, the number of features left varied from several to more than 20 features.

Next, feature selection was performed for each 1000-element group of features, using rfcv function in R's randomForest package. We applied log step, i.e. reducing a fixed proportion of variables at each run. The rfcv output is ranked by decreasing feature importance and the number of features left is chosen to minimize the out-of-bag (OOB, i.e. trees that did not have the validation examples in their bootstrap sample, see Sect. 4.1) error in CV. As a result, 8 $dFT330$ ms features were left, (OOB error 25.8%), 125 $FT1s$ features (OOB error 13.4%), 62 $FT330$ ms features (OOB error 20.4%). For *Basic* features we performed a similar procedure, but since in this case the log step removes a significant number of features, the output number of features was next increased and decreased with linear step, i.e. by a constant number of features, one by one. Initially 84 features were left (OOB error 6.11%), and after linear search 116 features were kept as a result. Altogether, $8 + 125 + 62 + 116 = 311$ features were left.

4   Classification

The experiments on the classification of the acquired data were performed using random forests (RF, [5]), deep learning (DL) architecture (neural network), and support vector machines (SVM), using R [18]. These classifiers were selected as they were used and performed well in other similar research, see [17,19].

The data were classified into three classes:

- deceleration, with speed decrease more than 5 km/h per second; this class represents fast deceleration, often happening when the driver must quickly reduce speed when speeding, before radar registers speed;
- stable speed, with speed change within 3 km/h range per second: from decrease below 1.5 km/h per second up to increase below 1.5 km/h per second;
- acceleration, with speed increase more than 2.5 km/h per second.

Dyno Data. The experiments were first performed on the data recorded at the dyno in Lublin. The data were labeled according to OBD data and other information recorded at the dyno. The speed ranges do not represent neighboring intervals, since we wanted to capture clear cases of intent acceleration or deceleration. The remaining data were not taken into account in our experiments. Altogether, we had 101 examples (1-s frames) representing deceleration, 423 examples for acceleration, and 579 examples for stable speed. Since deceleration was underrepresented, compared to the other classes, we also performed upsampling during training of classifiers (i.e. replicated examples, to match the number of examples in each fold with the biggest class), in order to balance classes.

On-Road Data. A pilot study on the on-road data was also performed. In this case a vehicle is moving wrt. the mic, and Doppler effect affects the recording. Since vehicles move fast, few audio frames are acquired in each move. Thus, we obtained a small set for pilot tests, with 18 examples for acceleration, 18 for deceleration, and 24 for stable speed. The examples were taken from twenty 1.66 s audio segments, i.e. three 1 s frames with 0.33 s hop size within each segment.

4.1   Classifiers

RF is a set of decision trees, and bias and correlations between the trees are minimized during the classifier construction. Each tree is built without pruning, using a different N-element bootstrap sample (i.e. obtained through drawing with replacement) of the N-element training set. For a K-element feature set, k features are randomly selected for each node of any tree ($k \ll K$, often $k = \sqrt{K}$). The best split on these k (called mtry in R) features is applied to split the data in the node. Gini impurity criterion is applied (minimized) to choose the split. This criterion measures how often an element would be incorrectly labeled, if random labeling an object according to the distribution of labels in the subset is applied. The forest of M trees is obtained by repeating this procedure M times. Classification using RF is performed by simple voting of all trees. RF yielded the best results in our experiments.

DL neural network in our experiments is a multi-layer feedforward neural net, with many hidden layers, with data standardization. Training is performed through back propagation with adaptive learning. We used R package h2o [15]; training parameters include large weight penalization and drop-out regularization (ignoring a random fraction of neuron inputs). In training, weights are iteratively updated in so-called epochs, with grid-search of the parameter space. DL yielded good results in our previous research on audio data for vehicles [19].

We also applied SVM classifiers, which look for a decision surface (hyperplane) maximizing the margin around the decision boundary. The training data points are called support vectors. SVM projects data into a higher dimensional space, using a kernel function, e.g. linear, quadratic, RBF (radial basis function). Each kernel has parameters which must be tuned to achieve good performance.

4.2   Classifier Tuning

Each of the classifiers we applied depends on some parameters. These parameters were tuned in the initial stage of the classification experiments.

The tuned parameters of RF were mtry, i.e. the number of features which are randomly selected for each node of any tree, nodesize, i.e. minimum size of leafs, and the number of trees. First, number of trees was estimated based on (averaged within 10 runs of RF training) OOB errors for consecutive number of trees, up to 500 trees. The lowest OOB error was found for about 420 trees. Next, grid search of RF parameters was performed using tune.randomForest from e1071 package in R [14], with the following search parameters: nodesize changing sequentially from 1 to 10, mtry = 17 (the square root of the feature set length), also doubled and halved (9 and 34), and the number of trees to grow 420 and nearest hundredths (400, 500). The output of the tuning is: 420 trees, mtry equal to the square root of the feature set length, and nodesize = 2. Additionally, we also built a model with nodesize = 1, as it is default setting in RF.

DL was trained using R's h2o package [15], with grid search for model tuning with the following search parameters:

- activation function: Rectifier, Tanh, Maxout, RectifierWithDropout, TanhWithDropout, MaxoutWithDropout.
- hidden layers - up to 4 layers: (20,20), (50,50), (30,30,30), (25,25,25,25).
- input dropout ratio $= (0,0.05)$.
- regularization: l1, l2, searched sequentially within (0,1e-4) with 1e-6 step.

The model found is: activation function Tanh, 2 hidden layers (50,50), input dropout ratio $= 0.05$, regularization l1 $= 6.3$E-5, l2 $= 5.1$E-5.

SVM was tuned using tune.svm from e1071 package in R [14]. Grid search of SVM parameters was done for linear, quadratic, and RBF kernel. Parameter c for the linear kernel was searched within $(2, 2^{10})$ range, with $c = 2$ found as best. For the quadratic and RBF kernels, γ was searched within $(2^{-5}, 2^5)$, and c within $(2, 2^{10})$, with $\gamma = 0.03125$ and $c = 2$ found as best in both cases.

5 Experiments

In the described experiments, we trained RF, DL and SVM classifiers to recognize acceleration, deceleration and stable speed for our data. Since the feature vector is large, we also performed feature selection. Initial tests (see Table 1) on dyno data, for *Basic* features only, were performed in 2 versions: with 10 best features kept, and with features of importance above 0.01 threshold, as 8-fold CV, where each fold represented data from one vehicle. The results for SVM without feature selection are low, but after feature selection the accuracy is better than for DL.

Table 1. Classification accuracy for speed changes

Classifier	RF	DL	SVM (RBF)
No feature selection	70.6%	60.9%	37.2%
Top 10 features	68.8%	58.7%	62.6%
Features above threshold	70.1%	55.5%	64.9%

We repeated these experiments after upsampling (Table 2). This increases accuracy in some cases, and our best result was for RF with feature selection above the threshold. However, upsampling decreases accuracy for DL with feature selection with top 10 features, which means that such a small amount of features is insufficient to classify such complex data.

The results show how difficult it is to classify such data, but looking into details (Tables 3 and 4) we can see that RF most often mistake stable speed with acceleration. Accelerating is a relatively slow process, so such mistakes are not surprising. Deceleration is mistaken with stable speed, but only engine braking deceleration was recorded, without applying brakes. Acceleration vs. deceleration

Table 2. Classification accuracy for speed changes after data balancing (upsampling)

Classifier	RF	DL	SVM (RBF)
No feature selection	66.9%	64.4%	36.4%
Top 10 features	67.3%	46.1%	60.7%
Features above threshold	72.6%	68.9%	65.7%

Table 3. Confusion matrix for RF (features above threshold, upsampling in training)

Class/Identified as:	Acceleration	Stable speed	Deceleration
Acceleration	300	118	5
Stable speed	106	447	26
Deceleration	0	47	54

Table 4. Confusion matrix for DL (features above threshold, upsampling in training)

Class/Identified as:	Acceleration	Stable speed	Deceleration
Acceleration	264	152	7
Stable speed	454	120	5
Deceleration	25	34	42

mistakes are rare, and deceleration was never mistaken for acceleration. The performance of DL is much worse for deceleration, and stable speed is much more often classified as acceleration than as stable speed.

Next, experiments were performed on full feature set, subjected to feature selection; we did not use the complete set in classification, as feature selection was assumed when creating the feature set. The results are similar to those from Table 1, but SVM with linear and quadratic kernels performed better than RBF:

- RF with `nodesize = 2`: 67.18%, with `nodesize = 1`: 66.46%,
- SVM with quadratic kernel 50.60%, linear 52.58%, RBF 41.16%,
- DL 57.30%.

5.1   Experiments on On-Road Data

For the purpose of our pilot study on the on-road data, it was necessary to take the Doppler effect into account. The Doppler shift changes frequencies in the spectrum, depending the moving object velocity, and the angle between the moving object with respect to the mic [3]. In order to obtain the training data with Doppler shift, we decided to use Adobe Audition CC [1] for adding Doppler effect to all dyno recordings. Only constant speed simulation was available in [1]. To obtain data approximating speed increase and decrease, we used frames representing different speed values when calculating dFT 330 ms and difference-based

Basic features. Lower speed in the starting frame and higher speed in the ending frame simulated increasing speed, whereas higher speed in the starting frame and lower speed in the ending frame simulated decreasing speed. The following speed pairs were used: 50 km/h and 55 km/h (or 55 + 50), 55 + 60 (60 + 55), 60 + 65 (65 + 60), 65 + 70 (70 + 65), and for deceleration also 60 + 50, 65 + 55, and 70 + 60. Two versions of Doppler shift were simulated: car moving left and right. Altogether, we had 101*14 = 1414 examples representing deceleration (101 original examples, 7 versions of deceleration in 2 directions), 423*8 = 3384 examples for acceleration (423 original examples, 4 versions of speed changes in 2 directions), and 579*10 = 5790 examples for stable speed (579 original examples and 5 speed values in 2 directions: 50, 55, 60, 65 and 70 km/h). For these data, the tuning of the classifiers was done again, with the following parameters obtained:

- RF: mtry = 248, 200 trees (error 1.8%), obtained using tuneRF from R's randomForest; nodesize = 1 as standard setting;
- DL: activation function Tanh, hidden layers (50,50), input dropout ratio = 0.05, regularization l1 = 6.3E-5 - as previously,
- SVM for linear kernel: $c = 2$ with lowest error (0.996724%), quadratic: $\gamma = 0.03125$, $c = 2$, RBF $\gamma = 0.03125$, $c = 2$.

Unfortunately, the results of classification were much lower than for the dyno data, i.e. 43.33% for DL, 40% for RF and SVM (RBF). We also performed tests on combined dyno and on-road recordings, divided into 10 folds, representing vehicles (for Toyota, both recordings were in one fold, other folds had either dyno or on-road data), obtaining 52.71% for DL, 54.43% for SVM (RBF), and 63.71% for RF, but we are aware that dyno-based folds raise the results.

Discussion. We are aware that on-road data differ from the dyno data, even after adding Doppler shift to the dyno recordings, and Doppler shift was only approximated for speed changes. Also, the on-road noises differ from the dyno noises, even for the same vehicle and tires. The road surface is more rough than the dyno, which is relatively smooth. There are also external noises around the road (birds etc.) and at the dyno (noises from outside when the door was open).

Time Series Approach. We also applied time series approach to the on-road data, namely shapelets [21], i.e. patterns in time domain. The experiments were performed on time-domain data directly, on MFCC coefficients observed in consequent frames (i.e. time series), and on spectrum magnitudes treated as a time series; these approaches are used in the literature. The data were limited to 1.5 s of each recording of a vehicle approaching the mic, i.e. with the Doppler shift in one direction only. The 0.1 s segment (4800 samples - we could not take 48000 for this algorithm) of one-channel audio data, taken from the center part of this 1.5 s, was used to find a shapelet, for each 1.5 recording, with results:

- time domain data: for 3 runs of the fast shapelet algorithm, the accuracy was 52.33%, 46.50%, and 44.92%,

- spectrum treated as a time series (magnitudes up to 150 Hz): 42.16%,
- MFCC coefficients as a time series: 38.77%,
- changes in time of spectrum bins; up to 150 Hz: 32.2% on average (17.5–52.5%), up to 500 Hz: 33% on average (12.5–65%),

The results are not satisfactory, even though these data were limited to relatively uniform (cars approaching, road data only).

Since these data differ from the data used earlier, we also performed experiments on such limited data (vehicles approaching the mic) after parameterization, using SVM, RF, and DL. Thus, we can compare results obtained for the same data. We also designed a new subset of features for this experiment.

Lines. We describe 3 strongest lines in 50–500 Hz spectrogram, present in the entire 1 s segment. No harmonicity is assumed. Each line is approximated with a linear function. The intercept, gradient, and quotient of the ending and starting frequencies for each line within the segment constitute a 9-feature group (*Lines*). No feature selection was applied. In classification with *Lines* only we obtained: DL 73.7%, RF 85.4%, SVM linear 73.6%, quadratic 62.3%, RBF 72.6%.

The results for the features from Sect. 3.1 and *Lines* for this limited dataset were as follows (results with *Lines* given if improved):

- DL, features from Sect. 3.1: 95.8%,
- RF, features from Sect. 3.1: 91.5% (94.3% with *Lines*),
- SVM, features from Sect. 3.1, linear kernel: 95.8%, quadratic: 92%, RBF: 61.3%.

These results show that carefully designed features can improve audio signal classification, and show the usefulness of our features.

6   Summary and Conclusions

In this paper, we aimed at automatic recognition of accelerating and decelerating of vehicles, as well as recognizing stable speed. Sound parameterization was devised to capture potential sound changes, but time domain was also directly applied using methods for time series. The data for the main experiments were recorded at the dyno, so we had access to the speed data and rpm (revolutions per minute) information. We also performed a pilot study with a small sample of on-road recordings. We used the dyno recordings with Doppler shift added, but these data were only approximation of Doppler effect, because of continuous speed changes in real world conditions.

The experiments on on-road data only were also performed using a time-series approach, on shorter audio segments, and compared with feature based approach. We designed an additional feature set for this experiment. The accuracy for these features only reached 85%, which shows that carefully designed feature set allows good classification.

We are planning to acquire more such data, use other classifiers, including classifier ensembles [16], and try other approaches to data balancing [4].

Acknowledgement. This work was partially supported by the Research Center of PJAIT, supported by the Ministry of Science and Higher Education in Poland.

References

1. Adobe. http://www.adobe.com/#
2. Ahn, I.-S., Bae, S.-G., Bae, M.-J.: Study on fault diagnosis of vehicles using the sound signal in audio signal processing. J. Eng. Technol. **3**, 89–95 (2015)
3. Berdnikova, J., Ruuben, T., Kozevnikov, V., Astapov, S.: Acoustic noise pattern detection and identification method in doppler system. Elektron. Elektrotech. **18**(8), 65–68 (2012)
4. Blaszczynski, J., Stefanowski, J.: Neighbourhood sampling in bagging for imbalanced data. Neurocomputing **150 A**, 184–203 (2015)
5. Breiman, L.: Random forests. Mach. Learn. **45**, 5–32 (2001)
6. Duarte, M.F., Hu, Y.H.: Vehicle classification in distributed sensor networks. J. Parallel Distrib. Comput. **64**, 826–838 (2004)
7. Erb, S.: Classification of vehicles based on acoustic features. Thesis, Graz University of Technology (2007)
8. EuroRAP. http://www.eurorap.pl/index.php
9. Hao, Y., Shokoohi-Yekta, M., Papageorgiou, G., Keogh, E.: Parameter-free audio motif discovery in large data archives. In: IEEE 13th International Conference on Data Mining, pp. 261–270 (2013)
10. Hastie, T., Tibshirani, R., Friedman, J.: The Elements of Statistical Learning. Data Mining, Inference, and Prediction. Springer Science+Business Media, LLC, Heidelberg (2009)
11. IRCAM. http://recherche.ircam.fr/anasyn/peeters/ARTICLES/Peeters_2003_cuidadoaudiofeatures.pdf
12. Moving Picture Experts Group. http://mpeg.chiariglione.org/standards/mpeg-7
13. Niewiadomy, D., Pelikant, A.: Implementation of MFCC vector generation in classification context. J. Appl. Comp. Sci. **16**(2), 55–65 (2008)
14. Package 'e1071'. https://cran.r-project.org/web/packages/e1071/e1071.pdf
15. Package 'h2o'. http://cran.r-project.org/web/packages/h2o/h2o.pdf
16. Rokach, L.: Ensemble-based classifiers. Artif. Intell. Rev. **33**(1), 1–39 (2010)
17. Siegel, J., Kumar, S., Ehrenberg, I., Sarma, S.: Engine misfire detection with pervasive mobile audio. In: Berendt, B., Bringmann, B., Fromont, É., Garriga, G., Miettinen, P., Tatti, N., Tresp, V. (eds.) ECML PKDD 2016, Part III. LNCS (LNAI), vol. 9853, pp. 226–241. Springer, Cham (2016). doi:10.1007/978-3-319-46131-1_26
18. The R Foundation. http://www.R-project.org
19. Wieczorkowska, A., Kubera, E., Słowik, T., Skrzypiec, K.: Spectral features for audio based vehicle identification. In: Ceci, M., Loglisci, C., Manco, G., Masciari, E., Ras, Z.W. (eds.) NFMCP 2015. LNCS (LNAI), vol. 9607, pp. 163–178. Springer, Cham (2016). doi:10.1007/978-3-319-39315-5_11
20. World Health Organization. http://www.who.int/mediacentre/factsheets/fs358/en/
21. Yeh, C.-C.M., Zhu, Y., Ulanova, L., Begum, N., Ding, Y., Dau, H.A., Silva, D.F., Mueen, A., Keogh, E.: Matrix profile I: all pairs similarity joins for time series: a unifying view that includes motifs, discords and shapelets. In: IEEE International Conference on Data Mining (2016)

A Statistical Approach to Speaker Identification in Forensic Phonetics

Fabio Leuzzi[1]([⊠]), Giovanni Tessitore[2], Stefano Delfino[2], Claudio Fusco[2],
Massimo Gneo[2], Gianpaolo Zambonini[2], and Stefano Ferilli[1]

[1] Dipartimento di Informatica, Università di Bari, Bari, Italy
{fabio.leuzzi,stefano.ferilli}@uniba.it
[2] Servizio Polizia Scientifica, Polizia di Stato, Rome, Italy
{giovanni.tessitore,stefano.delfino,claudio1.fusco,
massimo.gneo}@poliziadistato.it, gianpaolo.zambonini@interno.it

Abstract. Speaker identification can be summarized as the classification task that determines if two voices were spoken by the same person or not. It is a thoroughly studied topic, since it has applications in many fields. One is forensic phonetics, considered very hard since the expert has to face ambient noise, very short recordings, interference, loss of signal, and so on. For decades, these problems have been tackled by experts using their listening abilities, and each of them might represent a research area on its own. The use of semi-automatic techniques may represent a modern alternative to the subjective evaluation of experts, that may enforce fairness of the classification procedure. In a nutshell, we use the differences in speech of a set of different voices to build a population model, and the suspected person's voice to build a speaker model. The classification is carried out evaluating the similarity of a further speech sample (the evidence) with respect to the models. Preliminary evaluations shown that our approach reaches promising results.

1 Introduction

The *speaker identification* problem [11,15,16,35] can be cast as a classification task aimed at determining if two voices were spoken by the same person or not. In this broad sense, it has applications in many fields. In particular, it is a crucial task in forensics, where there is a need to determine the speaker in phone calls. This application domain adds further complexity to the task because calls are typically short in duration with poor quality, ambient noise, interference, loss of signal (in the case of mobile phones), and reduced bandwidth may yield dramatic consequences. Traditionally, the problem has been tackled leveraging abilities of human experts in evaluating the similarities between voices, or in finding peculiarities and defects that allow one to identify the speaker. However, this practice has its drawbacks, among which the limited capabilities of humans in considering complex mixes of parameters and their subjectivity in evaluation.

Nowadays, the most popular methods for speaker identification are the following: (1) listening based methods [24]; (2) spectrograms comparison

© Springer International Publishing AG 2017
A. Appice et al. (Eds.): NFMCP 2016, LNAI 10312, pp. 69–83, 2017.
DOI: 10.1007/978-3-319-61461-8_5

techniques [10,19]; (3) phonetic parameters analysis [1,12,27,37]; (4) automatic techniques [8,9]. In particular, the latter represents a modern alternative to overcome the subjective evaluation of experts, since it relies on algorithmic procedures to predict whether two voices come from the same speaker or not. So, it may ensure more fairness to the classification procedure.

The need of fairness is one of the main motivations for which this research field is so primary for judiciary contexts. Forensics aims to be a fair scientific support to the logical composition of crime events. In this case, such support regards the phone-speaker identification.

From a technical point of view, we can distinguish between *closed* tests (aimed at finding the speaker in a set of voices that surely includes a sample of the speaker's voice) and *open* tests (where this is not ensured). This paper deals with the open case, proposing a technique that uses the differences in speech of a set of different voices to build a population model, and the suspected person's voice to build a speaker model, and then carries out the classification by evaluating the similarity among these models and the anonymous voice. While a preliminary evaluation of this approach was presented in [34], this work aims at a specific analysis of results with respect to the feature selection perspective.

The remainder of this work will present some related work and preprocessing details in Sect. 2, then our approach follows in Sect. 3, after that experimental evaluation is reported in Sect. 4. Forensic results must be understandable to the Court, then Sect. 6 proposes a human understandable translation of the possible classification outcome. Finally, Sect. 7 will conclude with some considerations and future works.

2   Related Works, Background and Preprocessing

Different features may describe the sounds produced by the human vocal apparatus, depending on how it is classified. A first classification is between consonants and vowels. Consonants are produced by forcing air passage in the restricted vocal apparatus. They can be further divided in voiceless, if produced without vibration of vocal cords, or voiced, otherwise. Vowels are produced when the apparatus puts no obstacles, and the sound is determined by the position of tongue and lips. Specifically, they are a periodic signal produced by three factors: the periodic movement of vocal cords that produces the *fundamental frequency* (f_0 – related with the vocal tone of a person); the noise produced by the phonation; the modification of the sound caused by the sound expansion in the mouth. Such components make up the *frequency spectrum*. It is characterized by a sequence of peaks that change depending on the type of sound pronounced, a complex result of the cooperation of tongue, teeth, palate, lips, and so on. The frequency spectrum interacts with the harmonic structure of speech (integer multiples of the fundamental frequency). The harmonics near to the resonance frequency are called *formants*.

A spectrogram is a plot that represents the components of the sound in three dimensions: time (on the x axis), frequency (on the y axis) and intensity (represented using several color scales, here intensities of gray are exploited). The inner values are usually represented in Hertz. The lower frequency is known as *first formant* f_1, followed by the successive peaks named f_2, f_3, and so on. Generally, vowels are captured by f_1 and f_2, since the first formant indicates the vertical tongue movement (i.e., up or down), and the second indicates the horizontal tongue movement (i.e., back or forth). Furthermore, f_2 and f_3 may provide useful hints for the lips rounding. Formant frequencies are widely accepted features for use in forensic phonetics [18]. Several works are based on the study of f_0 only (e.g., [23]). Unfortunately, to date we cannot assert that voice is like a signature. So, in order to identify the speaker one needs as much information as possible, and it is questionable the fundamental frequency, alone, can be enough.

In order to overcome the uncertainty of the results using the fundamental frequency, [3] investigated the use of the first three formant frequencies and associated bandwidth. They are modeled using a multivariate Gaussian Mixture Model, in order to represent the vocal tract characteristics of the speaker, accounting for within-speaker variability. The results are expressed as a likelihood ratio, and highlight that since formants describe the cavity resonance, they are better suited for application in forensic speaker verification than Mel-Frequency Cepstral Coefficients (MFCC).

In [2], the authors focused on feature selection, investigating several ways to extract Cepstral Coefficients using the two major technologies for mobile communication (GSM and CDMA). Their approach uses the likelihood ratio to quantify the strength of speech evidence. The experiment highlighted the goodness of the MFCC, in spite of the outcomes obtained in [3]. They argue that such results are justified by the removal of the relevant information about the glottal shaping and lip radiation components due to the coding in mobile phone networks (both GSM and CDMA), that should make formant features useless.

In speaker verification task (i.e., the process of verifying the claimed identity of a speaker based on the speech signal), [5,25] create speakers model by measuring the fundamental frequency and formant frequencies of vowels (a, e, i, o), and estimating their distributions via Gaussian kernel density estimator. The long-term formant distributions are plotted and examined, accepting or rejecting the speaker. However, the authors pointed out that other information can be extracted from the shape of distributions. Likelihood Ratio [26] is exploited, like in this work, to evaluate the results in [5].

Our approach is text-independent, i.e. it tries to verify the identity without constraint on the speech content. We consider only a real-valued, limited, and continuous signal, i.e., a function that represents the proceeding of a given physical quantity (in our case, sound waves and their spectrum) over time. If a signal has period T (i.e., $x(t + T) = x(t)$), then the function is known when its proceeding in a range of length T is known. The inverse of T is the fundamental frequency $F = \frac{1}{T}$, measured in Hertz if time is expressed in seconds.

Conversely, formants are obtained from the signal spectrum. They are the resonance frequency measured where there is an energy peak in the sound produced by the air passage in the vocal apparatus, keeping into account absorptions due to the sound reflection. The fundamental frequency and the first three formants are the features of our speaker model.

The preprocessing step is carried out by a human operator that considers, for each word, only emphasized vowels, that are less affected by co-articulation and have a more constant signal than others. According to the literature (e.g., [5]), vowel U is not considered in this study. The human operator uses $Praat$[1], a software system able to show the graphical trend of the signal energy, allowing one to select the vowel to be analysed and to estimate the power spectrum. He selects the fundamental frequency f_0 using the $CEPSTRUM$ method (the result of the Fourier transformation applied to the signal spectrum measured in Decibel), and the formants f_1, f_2 and f_3 (i.e., the first three peaks of the frequency spectrum captured via Fast Fourier Transform). Subsequent formants cannot be detected, due to the poor signal quality.

Figure 1 shows the measurement of the formants of the first vowel E in the Italian word $gente$ (this particular GUI reports peaks of the spectrum, then the fundamental frequency f_0 is not reported in this Figure).

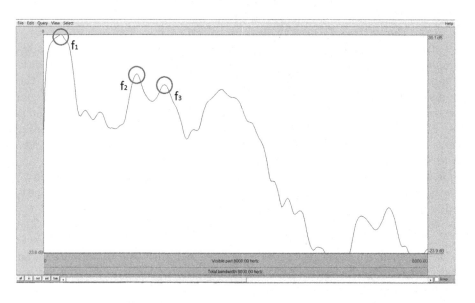

Fig. 1. Formants recognition using Praat.

3   Statistical Method Applied to the Recognition of the Talker - SMART

Suppose we have a distance measure. Then, we can describe the (possibly) large variability of voices among several speakers, as well as the small variability of

---

[1] www.fon.hum.uva.nl/praat/.

several dialogues of the same speaker. Such a variability can be represented estimating a distribution of the distances, making a model of the population diversities, together with a distribution of the diversities of the speech produced by a speaker in several contexts. Unfortunately, often there are not enough tracks of the same speaker to evaluate such distributions. So, a method to artificially populate a dataset related to the single speaker is needed. Bootstrap [28] resamples the dataset randomly picking whole records and repeating them. It cannot do otherwise, since the formants are related by complex relationships that impose to keep them together [30].

Missing Data. Often the recordings have poor quality, making hard the detection of some formants. We need to manage missing data. In order to face this necessity we adopted some policies. If the fundamental frequency is missing, the average of the known values is assigned to this cell; if a formant is missing, the cell is filled by its average conditioned by the values of the other formants, obtained via multiple regression; if the missing values are too many in a dimension (i.e. a feature), such dimension is removed completely, since estimations over few values are not reliable; it is noteworthy that first and second formants hardly are lacking. However, it might happen that there are no values for a vowel. In such a case the subject is pulled out from the dataset.

Speaker Representation. Given a generic speaker k, a generic vowel will be represented as $V_k \in \mathbb{R}^{N \times 4}$, where N is the number of that vowel instances, whereas 4 stands for the fundamental frequency and the three formants. Considering that we measured only the instances of vowels A, E, I and O, we have $V_{kj} \in \mathbb{R}^{N_{kj} \times 4}$ such that $j \in \{A, E, I, O\}$, and N_{kj} is the total amount of instances of the vowel j, speaker k. A speaker will be represented averaging the values over the columns, obtaining, for each V_{kj} a row vector $\bar{V}_{kj}$. Then:

$$\bar{S}_k = [\bar{V}_{kA}, \bar{V}_{kE}, \bar{V}_{kI}, \bar{V}_{kO}] \in \mathbb{R}^{1 \times 16}$$

where 16 is the total amount of vowels formants.

For the sake of completeness, we can give a fast look to an example of real data. Fixed the speaker k and the instance i_0 of the vowel A, we have:

$$V_{kA}(i_0) = [f_{00}, f_{01}, f_{02}, f_{03}]$$

an example record of which might be:

$$V_{kA}(i_0) = [129, 635, 1288, 2325]$$

Mahalanobis Distance and Statistical Distribution. Several measures have been investigated in [6, 7, 17, 29]. Summing up the results, these works shown the goodness of the Mahalanobis distance, that considers the position of the observations, it weights each observation with a coefficient extracted from the empirical covariance matrix. Such a matrix can be computed over the observed values, it represents the relationship between the features and shows how much a feature changes if related to the other ones.

A covariance matrix $\Sigma \in R^{16 \times 16}$ is computed over population matrix S, obtained chaining down the subjects in the population as shown in the following. In particular, recalling that each vowel of a speaker k is $V_{kj} \in \mathbb{R}^{N_{kj} \times 4}$ with $j \in \{A, E, I, O\}$, we expect different values for each N_{kj}, from which we can compute:

$$M_k = max_{j=\{A,E,I,O\}}(N_{kj})$$

The gap of each matrix that does not have M_k rows is filled. The instances are duplicated in the same order starting from the first, until the M_k number of rows is reached. The result, for each vowel, is a new matrix V'_{kj}. The representation of the speaker's data S_k will be:

$$S_k = [V'_{kA}, V'_{kE}, V'_{kI}, V'_{kO}] \in \mathbb{R}^{M_k \times 16}$$

where 16 is the total amount of vowels formants. Putting in a single matrix the set of available speakers, we obtain S.

At this point, given two subjects represented as $\bar{S}_i$ and $\bar{S}_j$, computed averaging column values of the respective matrices, the Mahalanobis distance $d(\cdot, \cdot)$ is:

$$d(\bar{S}_i, \bar{S}_j) = \sqrt{(\bar{S}_i - \bar{S}_j)\Sigma^{-1}(\bar{S}_i - \bar{S}_j)^T}$$

Now, suppose we have a pair of voices and we want to evaluate the possibility that they are produced by the same speaker or not. From a Bayesian point of view, we can introduce two statistical hypotheses to encode these possibilities. Say H_0 is the hypothesis that the two voices come from the same person (*accusatory hypothesis*):

$$P(H_0|d) = \frac{P(H_0)P(d|H_0)}{nf}$$

and H_1 is the hypothesis that the two voices do not come from the same person (*defensive hypothesis*):

$$P(H_1|d) = \frac{P(H_1)P(d|H_1)}{nf}$$

where nf is a normalization factor, which can be overlooked. We can combine them, obtaining:

$$\frac{P(H_0|d)}{P(H_1|d)} = \frac{P(H_0)}{P(H_1)} \frac{P(d|H_0)}{P(d|H_1)} \frac{nf}{nf} = \left(\frac{P(H_0)}{P(H_1)}\right) \cdot lr(d)$$

$$lr(d) = \frac{P(d|H_0)}{P(d|H_1)} = \frac{p_B(d)}{p_W(d)}$$

where $lr(d)$ denotes the likelihood ratio over d, $p_B(d)$ is the distribution of distance between the suspected speaker and the population (a.k.a. inter-distance), whereas $p_W(d)$ is the distribution of distance taken within different instances of the suspected speaker (a.k.a. intra-distance). Note that $p_B(d)$ and $p_W(d)$ are real valued functions of d. The strength of evidence is computed in $d(S_i, S_j)$ where S_i is the evidence speaker and S_j is the suspected speaker. Then, in our case, d is:

$$lr\big(d(S_i, S_j)\big) = \frac{P\big(d(S_i, S_j)|H_0\big)}{P\big(d(S_i, S_j)|H_1\big)} = \frac{p_B\big(d(S_i, S_j)\big)}{p_W\big(d(S_i, S_j)\big)}$$

Anyhow, the computation of $p_W(d)$ is not so direct, since often the sample is poor (just a few minutes of recording for the suspected person's voice). We need to refill the gap in order to have a number of simulated suspected-person's recordings comparable to the size of the population dataset. Then we recur to the bootstrap [28] procedure. It builds simulated registration using a random movement of the suspected person's data, generating as many suspected-person's samples as the subjects of the population. The Mahalanobis distance is computed for each pair of samples.

Estimating Speakers Distributions. In order to estimate $p_{B/W}(d)$ we exploit a semi-parametric kernel estimator method. Direct Plug-in Kernel [36] (as used in [7,17]), needs to estimate its smoothing parameter h, using $lsdpi(\cdot)$, shown in Algorithm 1. The semi-parametric kernel [4] is:

$$\tilde{p}_{B/W}(d) = \frac{1}{N} \sum_{j=1}^{N} \frac{1}{h} H\left(\frac{d - d_j}{h}\right)$$

where $\tilde{p}_{B/W}(d)$ denotes the model density of p_B or p_W, N is the size of the population, d_j is the distance between S_k and S_j, h is the smoothing parameter chosen via l-stage Direct Plug-in Kernel, $H(\cdot)$ is the kernel function (Gaussian in our case).

Such parameters ensure the satisfaction of:

$$H(\cdot) \geq 0$$

and

$$\int H(\cdot)du = 1$$

in this way the first formula will satisfy $\tilde{p}_{B/W}(d) \geq 0$ and $\int \tilde{p}_{B/W}(d)dx = 1$, as required for a function to be a probability density function.

4   Evaluation

We considered a dataset of Italian-male phone-call recordings, represented as described in Sect. 2 and made up as follows. $K = \{k_1, \cdots, k_i, \cdots, k_m\}$ is the set of pairs of same-speaker's recordings (in this experimental setting, recording 50 speakers twice). P is the set of single entries (they have not a paired recordings,

Algorithm 1. $lsdpi(\cdot) - l$-stage Direct Plug-in.

Input: Number of stages l, kernel function $K(\cdot)$ of order 2, a data sample X.
Output: Approximation of ψ_c.

$$\hat{\sigma} \leftarrow \sqrt{Var(X)}$$
$$c \leftarrow r + 2l$$
$$\psi_c \leftarrow \frac{(-1)^{\frac{c}{2}} c!}{(2\hat{\sigma})^{c+1}(\frac{c}{2})!\sqrt{\pi}}$$
$$c \leftarrow c - 1$$
while $c \geq 1$ **do**
$$g \leftarrow \left[\frac{-2K^c 0}{\mu_2(K)\psi_{c+2}n} \right]^{\frac{1}{2c+5}}$$
$$\psi_c \leftarrow n^{-1} \sum_{i=1}^{n} \sum_{j=1}^{n} K_g^r(X_i - X_j)$$
$$c \leftarrow c - 2$$
end while
return ψ_c

so they can be used only as negative examples – in this experimental setting, we evaluated 350 single entries). So, we have just K positive test, while the number of negative tests will be:

$$nt(K, P) = \frac{P(P-1)}{2} + 2KP + \left(4\frac{K(K-1)}{2} \right)$$

In this experimental setting, $nt(50, 350) = 100.975$. Our evaluation has a two-fold objective: on the one hand, understanding the performance, on the other, finding the set of formants that best represent a voice signature.

Table 1. Feature-subset performances

Tested features	EER	AUC
f_0, f_1, f_2, f_3	0.07692	0.98620
f_1, f_2, f_3	0.07692	0.98108
f_0, f_2, f_3	0.05406	**0.99295**
f_0, f_1, f_3	0.07658	0.98587
f_0, f_1, f_2	0.07182	0.98295
f_0, f_1	0.11538	0.96489
f_0, f_2	0.08800	0.97306
f_0, f_3	**0.03980**	0.98952
f_1, f_2	0.09615	0.97104
f_1, f_3	0.11022	0.96991
f_2, f_3	0.07692	0.97328

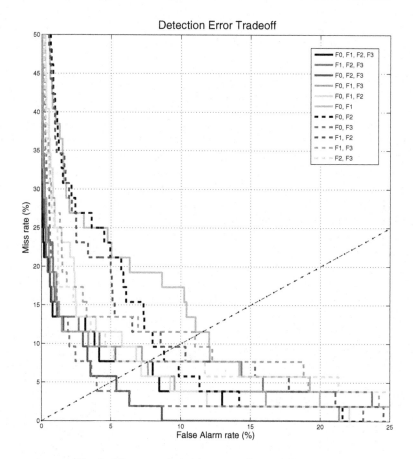

Fig. 2. Diagram of the feature-subset performances

The Likelihood-Ratio $lr(\cdot)$, reported in Sect. 3, expresses how many times more likely we can observe distance d between unknown and evidence voices under the accusatory hypothesis than the defensive hypothesis. It has been used to build the graph in Fig. 2, that shows Detection Error Curve for each subset of features. Table 1 shows the Equal Error Rate (EER) and the Area Under the ROC Curve (AUC) for each subset. The former value is useful to balance the misclassification types, whereas the latter is used to identify the subset that makes less mistakes. The best EER in Table 1 is the curve nearest to zero in Fig. 2, i.e. the subset $\{f_0, f_3\}$, whereas the best AUC in Table 1 is the curve that goes faster to zero in both dimensions in Fig. 2, i.e. $\{f_0, f_2, f_3\}$.

Since both include f_0 and f_3, we should comment the role of f_2. Examining the subset $\{f_0, f_2, f_3\}$ in Fig. 2 we can see that the *false alarm rate* (the worst justice mistake) goes to zero faster than others. Looking at the values in Table 1, the EER of $\{f_0, f_2, f_3\}$ is greater than the EER of $\{f_0, f_3\}$ just a little bit with respect the trend of growth of the EER in general. Furthermore, the subset $\{f_0, f_2, f_3\}$ gives the maximum value of AUC, denoting the smallest error area.

5   SMART: A Particular Case of a Biometric System in the Bayesian Framework

In Sect. 3 we have described step-by-step how the system SMART works. Anyway, it can be seen as a particular case of a more general *Biometric system* used to compute the strength of evidence in terms of Likelihood Ratio in the Bayesian Framework. Let us give a look to Fig. 3. Say E_1 is the crime-scene evidence, and E_0 is the suspected-person's one. For the sake of clarity, example of such pairs could be DNAs, fingerprints, Photos (one from a video-surveillance system that recorded the crime, and the other from suspected-person's); audio tracks, as in our case; and so on.

Whatever is the evidence type, the objective is to establish the strength of evidence that E_0 and E_1 belong to the same person versus the hypothesis that they come from different persons. This objective can be framed in the Bayesian framework as introduced in Sect. 3, in which the Likelihood Ratio is the strength-of-evidence measure.

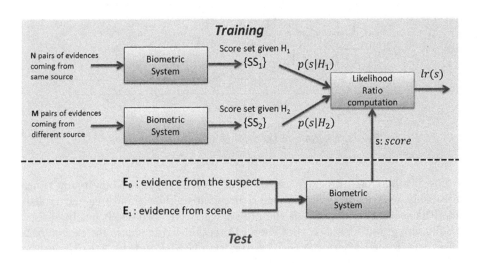

Fig. 3. Biometric Bayesian framework

In order to estimate the Likelihood Ratio all we need is a "black box" able to compute the similarity "score", between two evidences and a database containing both pairs of evidences coming from same persons and pairs coming from different ones. Note that in SMART such black box is simply the Mahalanobis distance between a pair of speakers and the score that can be obtained by the inverse of the distance. Going on, the black box *Likelihood Ratio computation* works in two steps:

- **the training phase**, in which two sets of scores are computed from pairs of evidences coming from same (i.e. SS_1) and different (i.e. SS_2) source(s) contained in the database. The sets of scores SS_1 and SS_2 are used to infer the score distributions given the accusatory (i.e. H_1) and the defense (i.e. H_2) hypothesis respectively. In SMART, the set SS_1 is obtained using the bootstrap, whereas the score distributions are computed using a semi-parametric kernel estimator method. After that SMART estimates the Likelihood-Ratio function (i.e. $lr(\cdot)$, or $lr(1/d)$ as defined in Sect. 3). For the computation of SS_2 there are two different approaches: suspect anchored and suspect independent. The former computes SS_2 as the set of scores between the suspected-person's evidence and each other evidence belonging to the database. The latter approach computes SS_2 as the set of scores between all possible (different) pairs of evidences stored in the database. SMART is a suspect-anchored approach;
- **the test phase**, in which the Biometric-system box is used to compute the score between E_0 and E_1. The resulting score value is exploited to obtain the final Likelihood Ratio.

6 Presenting Likelihood Ratio to the Court

Noteworthy, the bootstrap makes our approach non-deterministic, for which the evaluation between speakers (S_1, S_2) is different from (S_2, S_1). There is no theoretical reason to apply the bootstrap to the suspected speaker instead of anonymous one; given that the same classification is expected applying the approach in both directions. Anyway, from a practical point of view, suspected-person's data are often more rich than anonymous, since when the suspects arise, there is enough time to organize the activities in order to record as much dialogues as possible. This is the reason for which the only reliable classification is carried out applying the bootstrap to the suspected-person's data.

We recall that the value of Likelihood Ratio $lr(\cdot)$ quantifies the strength of the evidence. This values must be presented to the Court, then the *European Network of Forensic Science Institutes* (a.k.a. ENFSI) provided detailed guidelines for this purpose[2].

In order to cope with the great amount of different applications of $lr(\cdot)$, its logarithm is commonly used, known as Log-Likelihood Ratio. Given two speakers S_i and S_j, and a distance $d(S_i, S_j)$ on which $lr(d)$ is computed, the Log-Likelihood Ratio function $llr(d)$ is:

$$llr(d) = Log_{10}(lr(d))$$

Tables 2 and 3 show the ranges used to evaluate this proposal. For the sake of completeness, corresponding Log-Likelihood Ratio is reported, given that it

---

[2] With the financial support of the Prevention and Fight against Crime Program of the European Union European Commission - Directorate - General Justice, Freedom and Security. A project funded by the EU ISEC 2010. Agreement number: HOME/2010/ISEC/MO/4000001759.

Table 2. $lr(\cdot)$ values supporting the prosecution hypothesis

$lr(\cdot)$	$llr(\cdot)$	Typical translation
>10000	>4	Very strong evidence to support
1000 to 10000	3 to 4	Strong evidence to support
100 to 1000	2 to 3	Moderately strong evidence to support
10 to 100	1 to 2	Moderately evidence to support
1 to 10	0 to 1	Limited evidence to support

Table 3. $lr(\cdot)$ values supporting the defense hypothesis

$lr(\cdot)$	$llr(\cdot)$	Typical translation
<0.0001	<−4	Very strong evidence to support
0.001 to 0.0001	−3 to −4	Strong evidence to support
0.01 to 0.001	−2 to −3	Moderately strong evidence to support
0.1 to 0.01	−1 to −2	Moderately evidence to support
1 to 0.1	0 to −1	Limited evidence to support

is easy to use and widely adopted. Since the numeric form of a $lr(\cdot)$ may not be readily interpretable to the Court, the last column reports translations into verbal scale, that prosecutor (Table 2) and defender (Table 3) lawyers can use to present the classification result to the Court.

7   Conclusion

This work presented an approach to Speaker Identification that models the speaker via fundamental frequency and formant features. Distances among these descriptions have been computed using the Mahalanobis distance, in order to model the typical distance in speech among several speakers. Such a model has been obtained estimating the distributions of the differences. In particular, both the set of different speakers and the set of tracks recorded from the same speaker have been modeled, in order to obtain comparable models useful to decide if a novel speaker description is nearest to the unknown speaker model or it is nearest to the population model.

The interpretation of SMART as a general Biometric System, working in the Bayesian framework, provides novel insights for future developments and tests. For instance, one could try to assess how the performance change varying score functions, after that the investigation could follow with the comparison of the outcomes using suspect anchored or suspect independent approach. Moreover, we recall that in forensics it is mandatory to have a system with good discrimination ability (generally verified by AUC and/or EER), but it is mandatory also to have a reliable Likelihood Ratio, making fundamental an investigation about the use of Cost Likelihood Ratio.

Other future works will be focused on clustering speaker description via unsupervised techniques, in order to understand whether formant features are enough to obtain clusters representing italian dialects. Moreover in forensics the discrimination of the model measured by AUC and EER does not suffice to measure the reliability of the computed Likelihood Ratio. For example it is mandatory that the system does not give high positive/negative Log-Likelihood Ratio for the wrong hypothesis. To this aim other error functions, such as Cost-Likelihood Ratio error function, will be investigated in future works.

References

1. Federico, A., Ibba, G., Paoloni, A.: A new automated method for reliable speaker identification and verification over telephone channel. In: ICASSP, p. 1457 (1987)
2. Alzqhoul, E.A.S., Nair, B.B.T., Guillemin, B.J.: Comparison between speech parameters for forensic voice comparison using mobile phone speech. In: Speech Science and Conference 2014 (2014)
3. Becker, T., Jessen, M., Grigoras, C.: Forensic speaker verification using formant features and Gaussian mixture models. In: INTERSPEECH, pp. 1505–1508. ISCA (2008)
4. Bishop, C.M.: Neural Networks for Pattern Recognition. Oxford University Press Inc., New York (1995)
5. Grigoras, C.: Forensic voice analysis based on long term formant distributions. In: 4th European Academy of Forensic Science Conference (2006)
6. Calvani, F.: Il problema dell'errore di assegnazione nel riconoscimento del parlatore. Tesi di laurea in Matematica, Universit Tor Vergata di Roma (1996)
7. Calvani, F.: Analisi critica di metodi per la classificazione del parlatore nelle scienze forensi. Tesi di laurea in Matematica, Universit Tor Vergata di Roma (1998)
8. Drygajlo, D.: Forensic automatic speaker recognition. IEEE Sig. Process. Mag. **24**, 132–135 (2007)
9. Drygajlo, A., Meuwly, D., Alexander, A.: Statistical methods and Bayesian interpretation of evidence in forensic automatic speaker recognition. In: EUROSPEECH 2003, Geneva, Switzerland, pp. 689–692 (2003)
10. Koenig, B.E.: Selected topics in forensic voice identification. Crime Lab. Dig. **20**(4), 78–81 (1993)
11. Mathan, L., Bimbot, F., Magrin-Chagnolleau, I.: Second-order statistical measures for text-independent speaker identification. Speech Commun. **17**, 177–192 (1995)
12. Falcone, M., Paoloni, A., De Sario, N.: IDEM: a software tool to study vowel formant in speaker identification. In: Proceedings of the ICPHS 1995, Stockholm, vol. 3, pp. 294–297 (1995)
13. Ferilli, S., Leuzzi, F., Rotella, F.: Cooperating techniques for extracting conceptual taxonomies from text. In: Proceedings of the Workshop on Mining Complex Patterns at AI*IA XIIth Conference (2011)
14. Ferilli, S., Leuzzi, F., Rotella, F.: A run length smoothing-based algorithm for non-Manhattan document segmentation. In: Proceedings of Convegno del Gruppo Italiano Ricercatori in Pattern Recognition (2012)
15. Furui, S.: Digital Speech Processing, Synthesis and Recognition. Marcel Dekker Inc., New York (1989)
16. Paoloni, A., Ibba, G.: Analisi delle voci: il parlatore ignoto. Poste e Telecomunicazioni, pp. 14–25 (1993)

17. Ghizzoni, A.: Il problema dell'identificazione del parlatore nelle scienze forensi: modelli, metodi di classificazione e analisi dei dati. Tesi di laurea in Matematica, Universit Tor Vergata di Roma (1999)

18. Grimaldi, M., dApolito, S., Gili Fivela, B., Sigona, F.: Illusione e scienza nella fonetica forense: una sintesi. Mondo digitale (2014)

19. Kersta, L.J.: Voiceprint identification. Nature **196**, 1253–1257 (1962)

20. Leuzzi, F., Ferilli, S., Rotella, F.: Improving robustness and flexibility of concept taxonomy learning from text. In: Appice, A., Ceci, M., Loglisci, C., Manco, G., Masciari, E., Ras, Z.W. (eds.) NFMCP 2012. LNCS, vol. 7765, pp. 170–184. Springer, Heidelberg (2013). doi:10.1007/978-3-642-37382-4_12

21. Leuzzi, F., Ferilli, S., Rotella, F.: ConNeKTion: a tool for handling conceptual graphs automatically extracted from text. In: Catarci, T., Ferro, N., Poggi, A. (eds.) IRCDL 2013. CCIS, vol. 385, pp. 93–104. Springer, Heidelberg (2014). doi:10. 1007/978-3-642-54347-0_11

22. Leuzzi, F., Ferilli, S., Rotella, F.: A relational unsupervised approach to author identification. In: Appice, A., Ceci, M., Loglisci, C., Manco, G., Masciari, E., Ras, Z.W. (eds.) NFMCP 2013. LNCS, vol. 8399, pp. 214–228. Springer, Cham (2014). doi:10.1007/978-3-319-08407-7_14

23. Lindh, J.: Preliminary F0 statistics and forensic phonetics. In: Lindh, J., Eriksson, A. (eds.) Annual Conference of IAFPA, Department of Linguistics, Gteborg University (2006)

24. Nolan, F.: Speaker recognition and forensic phonetics. In: The Handbook of Phonetic Sciences (1997)

25. Nolan, F., Grigoras, C.: A case for formant analysis in forensic speaker identification. Int. J. Speech Lang. Law **12**(2), 143 (2005)

26. Rose, P.: Forensic Speaker Identification. Taylor & Francis London, New York (2002)

27. Paoloni, A., Falcone, M., Federico, A.: The parametric approach in forensic speaker recognition. In: Proceedings of COST 250 Workshop on Speaker Recognition by Man and Machine: Directions for Forensic Applications, pp. 45–51 (1998)

28. Rosati, F.: Sperimentazione del metodo bootstrap nel problema del riconoscimento del parlatore. Tesi di laurea in Matematica, Universit Tor Vergata di Roma (2001)

29. Rossi, C.: Il problema di decisione dell'identificazione del parlatore. Caratterizzazione del parlatore, pp. 173–176 (1996)

30. Rossi, C.: Classification and decision making in forensic sciences: the speaker identification problem. In: Rizzi, A., Vichi, M., Bock, H. (eds.) Advances in Data Sciences and Calssification, pp. 647–654. Springer, Heidelberg (1998). doi:10.1007/ 978-3-642-72253-0_88

31. Rotella, F., Ferilli, S., Leuzzi, F.: An approach to automated learning of conceptual graphs from text. In: Ali, M., Bosse, T., Hindriks, K.V., Hoogendoorn, M., Jonker, C.M., Treur, J. (eds.) IEA/AIE 2013. LNCS, vol. 7906, pp. 341–350. Springer, Heidelberg (2013). doi:10.1007/978-3-642-38577-3_35

32. Rotella, F., Ferilli, S., Leuzzi, F.: A domain based approach to information retrieval in digital libraries. In: Agosti, M., Esposito, F., Ferilli, S., Ferro, N. (eds.) IRCDL 2012. CCIS, vol. 354, pp. 129–140. Springer, Heidelberg (2013). doi:10.1007/ 978-3-642-35834-0_14

33. Rotella, F., Leuzzi, F., Ferilli, S.: Learning and exploiting concept networks with connektion. Appl. Intell. **42**(1), 87–111 (2015)

34. Forte, A., Rossi, C., Bove, T., Giua, P.E.: Un metodo statistico per il riconoscimento del parlatore basato sull'analisi delle formanti. Statistica LXII, 177–192

35. Furui, S., Matsui, T.: Adaptation of tied mixture based phoneme models for text-prompted speaker verification. In: ICASSP, pp. 125–128 (1994)
36. Wand, M.P., Jones, M.C.: Kernel Smoothing. Monographs on Statistics and Applied Probability. Chapman & Hall/CRC, Boca Raton, London, New York (1995)
37. Wolf, J.J.: Efficient acoustic parameters for speaker recognition. J. Acoust. Soc. Am. **51**(6), 2044–2056 (1972)

Increasing the Interpretability of Rules Induced from Imbalanced Data by Using Bayesian Confirmation Measures

Krystyna Napierała[1,2], Jerzy Stefanowski[1], and Izabela Szczęch[1(✉)]

[1] Institute of Computing Science, Poznań University of Technology,
60-965 Poznań, Poland
`izabela.szczech@cs.put.poznan.pl`
[2] DATAX Sp. z o.o., 53-609 Wroclaw, Poland

Abstract. Approaches to support an interpretation of rules induced from imbalanced data are discussed. In this paper, the rule learning algorithm BRACID dedicated to class imbalance is considered. As it may induce too many rules, which hinders their interpretation, their filtering is applied. We introduce three different strategies, which aim at selecting rules having good descriptive characteristics. The strategies are based on combining Bayesian confirmation measures with rule support, which have not yet been studied in the class imbalance context. Experimental results show that these strategies reduce the number of rules and improve values of rule interestingness measures at the same time, without considerable losses of prediction abilities, especially for the minority class.

Keywords: Bayesian confirmation measures · Interpretability of rules · Class imbalance · Rule post-pruning

1 Introduction

Learning classification rules is one of mature and well studied tasks in machine learning. The popularity of rules comes from the fact that they directly provide a symbolic representation of knowledge discovered from data, which is more comprehensible and human-readable than other representations [5]. Many various algorithms for inducing rules have been already introduced (for their review see, e.g. [5]). Nevertheless, such aspects of data complexity as *class imbalance* still constitute difficulties [11]. The majority of standard rule algorithms are biased towards the majority classes and tend to neglect the minority class. Two kinds of reasons for poor performance of rule based classifiers for imbalanced data are usually pointed out – algorithmic and data level ones [11,16].

Some extensions of rule classifiers for class imbalances have been already proposed, for their review see [16]. However, most of them address only a single or at most a few of algorithmic or data-related factors. In [16] we introduced a new rule induction algorithm, called BRACID (the acronym of Bottom-up induction of Rules And Cases for Imbalanced Data), which attempts to deal with

A. Appice et al. (Eds.): NFMCP 2016, LNAI 10312, pp. 84–98, 2017.
DOI: 10.1007/978-3-319-61461-8_6

more of the aforementioned factors. The previous comparative experiments have clearly demonstrated that the rule classifier induced by BRACID significantly outperformed other rule classifiers generated by the best, standard rule learning algorithms as well as the rule extensions specialized to class imbalances, with respect to predictive measures [16]. On the other hand, BRACID may generate too many rules (see also experiments in Sect. 6). As it restricts human experts' abilities to analyze or interpret the rules, we are looking for a post-processing approach that could identify the most valuable rules. The first attempt, recently undertaken in [18], has shown that it is possible to select rules characterized by high supports and still leading to sufficient predictive performance.

Nevertheless, focusing attention on the most interesting rules should also take into account other characteristics than simply the rule support. In particular, it is important not to neglect the descriptive abilities of rules, which are often overwhelmed by the need to increase the predictive performance. Note that the predictive and descriptive aspects often stand in opposition to each other [13,20]. However, when human experts seek for a compact knowledge representation, improving the interpretability of each single rule can even justify some loses on the predictive performance.

Establishing when rules are interesting to users touches both subjective and objective aspects [4]. In this paper we follow the latter aspect and consider *rule interestingness measures* which are often applied to filter the set of rules [7,15]. They are calculated from learning data and aim at quantifying the relationship between a rule's premise and its conclusion. A particular group of these measures, called *Bayesian confirmation measures*, is well suited for supporting rule interpretability, as it focuses on advancing rules for which the probability of the conclusion given the premise is greater than the genuine probability of the conclusion itself [3,10]. In other words, confirmation measures promote rules, in which the value that the premise adds to conclusion is considerably high.

Although the concept of confirmation has been firstly considered by philosophers of science in a very different context (see e.g. [2,3,19]), it has been adopted to rule interestingness measures, mainly for filtering association rules [8]. Nevertheless, these measures have not been considered for imbalanced data yet. Their application should turn out to be particularly useful in the context of imbalance since considering the probability of each conclusion separately, as done by confirmation measures, would be related to imbalance ratios.

For the purpose of this paper we focus on two particular confirmation measures called S [2] and N [19]. We have chosen them from a wider collection of confirmation measures discussed in the literature because of the desired properties that they possess [9,10]. In our opinion, these measures satisfy properties that should influence the interpretability of rules [10].

The main aim of this paper is to introduce an approach that uses confirmation measures S and N to post-prune rules induced by BRACID. We focus this study on BRACID only, as experiments [16] have shown that it outperformed other best rule based classifiers over a large collection of imbalanced data. The new approach should reduce the number of its rules while improving values of rule interestingness measures at the same time, especially for the minority class.

The paper is organized as follows. Firstly, we briefly review related works in Sect. 2. Section 3 introduces the concept of Bayesian confirmation and defines two particularly valuable representatives of confirmation measures: measures S and N. The algorithm used for rule induction, called BRACID, is summarized in Sect. 4. The three new rule filtering strategies are introduced in Sect. 5. Their usefulness to improve BRACID rules is evaluated in several experiments, which are described in Sect. 6. The experimental results are discussed in Sect. 7. In the final section we draw lines of future research.

2   Related Works on Rule Evaluation and Filtering

Many algorithms for constructing rule based classifiers employ rule pruning. The representative approaches are Grow, IREP or RIPPER; for their review, see Chap. 9 in [5]. However, these approaches follow the *classification perspective of rule induction* and pruning is oriented toward good predictive ability of the complete set of rules. Other objectives are stated in the *descriptive knowledge discovery* which aims at discovering from data information patterns and regularities (or sometimes exceptions) which are potentially *interesting* and *useful* to different kinds of users [20].

The descriptive rule discovery perspective, which is considered in this paper, requires other algorithmic strategies than in the classification perspective, e.g., classification versions of association rules, richer sets of satisfactory rules [20] or rule representations of subgroup discovery [6]. However, these algorithms often generate a too high number of rules which makes it impossible for users or domain expert to inspect them. Thus, users lose the opportunity to interpret the results, find interesting rules or to further modify them to have a more accurate classifier [12].

To help the user find relevant knowledge inside huge rule sets, the *rule interestingness measures* have been proposed (for their review see [7,15]). They are divided into two categories: *subjective* (user-oriented) and *objective* (data-oriented) ones. The subjective measures take into account the user's goals, background knowledge or his belief on the data domain [4]. Objective measures are those that are not application- or user-specific and depend only on raw data. Many of them are defined on the basis of contingency tables summarizing the data set (see the next section). *Support* and *confidence* are the most universal interestingness measures which are often applied in the process of rule generation (e.g., Apriori search for association rules) and sometimes in post-filtering [1]. Although they are so popular, other measures could be better suited to deal with larger sets of rules and to select the most relevant (i.e. interesting) candidates. Numerous rule interestingness measures have been proposed (lists can be found for example in [7,12,13,15]) and choosing the best one for a given problem is not a trivial task.

In general, the interestingness measures are used to assess, rank (sort) and filter the rules according to various points of view [7]. For these aims, the experts either select some single measures or consider their aggregated, more complex versions. For instance, [13] describes a case study in which several measures have been used, and the results were interpreted by an expert with a recommendation to use a weighted relative accuracy. Another, more multiple-criteria analysis has been advocated by Bayardo and Agrawal, who proposed to analyze partial ordering of the rules (instead of the typical total ordering of rules) according to different interestingness measures [1]. The authors of [12], on the other hand, discussed other related proposals and proposed a subset of measures based on specialist's preferences; see also [14]. The authors of [9] analyzed properties of the interestingness measures and showed that some measures may be preferred to others. Furthermore, other researchers looked for concise representations (e.g. closed items in associations), rule summaries, grouping of similar rules (with respect to rule condition parts or to subsets of covered examples), or developed interactive visualization tools.

Nevertheless, the choice of the interestingness measures still depends on the expert's preferences and the problem at hand. In this paper, following motivations presented in Sect. 1, we direct our interest to a particular class of measures based on Bayesian confirmation. Although they have been recently used to filter association rules [8], they have not been considered for classification rules in the class imbalanced tasks.

3 Bayesian Confirmation Measures

To present Bayesian confirmation measures the basic notation is introduced. Rules are consequence relations represented as *IF (condition part) THEN (target class)*, where a condition part (premise) is a conjunction of elementary tests on values of attributes characterizing learning examples and a target class points to one of the predefined values of the decision attribute (represented in a rule conclusion). For simplicity, rules will be denoted as $E \rightarrow H$ or simpler as R.

Interestingness measures quantify the relationship between E and H, and are usually defined as functions of four non-negative values that can be gathered in a 2×2 contingency table (see Table 1). For a particular data set, a is the number of objects that satisfy both the rule's premise and its conclusion, b is the number of learning examples for which only H is satisfied, etc. For instance, the support of $E \rightarrow H$ rule is defined as $sup(H, E) = a$ and its confidence as $conf(H, E) = a/(a+c)$. Note that a, b, c and d can also be regarded as frequencies for estimating probabilities: e.g. $P(E) = (a + c)/n$ or $P(H) = (a + b)/n$.

Table 1. An exemplary contingency table of the rule's premise and conclusion

	H	$\neg H$	Σ
E	a	c	$a + c$
$\neg E$	b	d	$b + d$
Σ	$a + b$	$c + d$	n

Among many interestingness measures, we drew our attention to a particular group of *Bayesian confirmation measures* (or simply *confirmation measures*). All those measures are characterized by a feature called *property of Bayesian confirmation*, which requires that an interestingness measure $c(H, E)$ obtains: positive values when $P(H|E) > P(H)$; 0 when $P(H|E) = P(H)$; and negative values when $P(H|E) < P(H)$.

Thus, confirmation measures are designed to depict simply through their scale the confirmatory, neutral or disconfirmatory impact of the rule's premise on its conclusion. Confirmation, interpreted as an increase in the probability of the conclusion H provided by the premise E, is a desirable situation. Let us stress that basic interestingness measures such as support or confidence do not possess the property of confirmation and thus, their utility is lower for the descriptive perspective of knowledge discovery.

The difference of semantics and utility of confidence on one hand, and measure $S(H, E)$ (defined below in Eq. 1) being a representative of confirmation measures on the other hand, can be shown on the following illustrative example. Consider the possible result of rolling a dice: 1, 2, 3, 4, 5, 6 points, and let the conclusion $H =$ "*the result is divisible by 2*". Given two different potential rule premises:

$E_1 =$ "*the result is a number from a set {1, 2, 3}*",
$E_2 =$ "*the result is a number from a set {2, 3, 4}*"

we get, respectively: $conf(H, E_1) = 1/3$, $S(H, E_1) = -1/3$ and $conf(H, E_2) = 2/3$, $S(H, E_2) = 1/3$. This example clearly shows that the values of confirmation measures have a more useful interpretation than confidence. In particular, in the case of rule $E_1 \rightarrow H$, the premise actually disconfirms the conclusion as it reduces the probability of conclusion H from $1/2 = P(H)$ to $1/3 = P(H|E_1) = conf(H, E_1)$. This fact is expressed by a negative value of confirmation measure $S(H, E)$ (and in fact any confirmation measure), but it cannot be concluded by observing only the value of confidence.

Note that the property of confirmation leaves plenty of space for defining various, non-equivalent confirmation measures (for their review see [3,9]). To guide the user towards the measures that reflect his expectations, researchers proposed special properties of confirmation measures. These properties express requirements for a measure behavior in certain situations. Taking into account possession of desirable properties, we focus our further interest only on two representatives of confirmation measures. The chosen measures $S(H, E)$ [2] and $N(H, E)$ [19], both ranging from -1 (showing complete disconfirmation) to $+1$ (showing complete confirmation), are defined as:

$$S(H, E) = P(H|E) - P(H|\neg E) = \frac{a}{a + c} - \frac{b}{b + d}, \tag{1}$$

$$N(H, E) = P(E|H) - P(E|\neg H) = \frac{a}{a + b} - \frac{c}{c + d}. \tag{2}$$

Among properties that valuable confirmation measures should satisfy let us mention property of monotonicity M [10] and property of *maximality/minimality* [8]. Monotonicity M favors measures that are non-decreasing with respect to a and d, and non-increasing with respect to b and c. It is intuitively clear that we would like higher values of measures for rules that are supported by a greater number of positive examples (i.e. increase of a), and exactly the opposite when the number of counter-examples grows (i.e. increase of c). The property of *maximality/minimality* on the other hand, requires that a measure obtains its maximal value if and only if $b = c = 0$, and its minimal values if and only if $a = d = 0$. It is thus a property concentrated on the behavior of measures in the extreme cases. It was verified in [9,10] that the measures $S(H, E)$ and $N(H, E)$ are among few confirmation measures that satisfy both monotonicity M and *maximality/minimality*.

We have focused our study on those two measures also because the interpretation of their definitions is rather straightforward (contrary to some other confirmation measures possessing M and *maximality/minimality* e.g. measure $c_3(H, E)$ [9][1]). Measure $S(H, E)$ expresses how much more probable is H with E rather than with $\neg E$. Following some medical examples, e.g. if some symptoms occur then a certain disease is diagnosed, we could say that measure $S(H, E)$ assesses how much more probable becomes the disease when we know that the symptoms occurred (instead of knowing that the symptoms did not occur). In case of measure $N(H, E)$, we would say that is expresses how much more probable are some symptoms for a certain disease than for a case when the disease is excluded (does not occur). Measures $S(H, E)$ and $N(H, E)$ are thus somewhat complementary, as they look at rules from different perspectives: that of the rule's premise and that of the rule's conclusion.

Summing up, taking into account possession of desirable properties and interpretation of the measures' definitions, this study focuses only on application of confirmation measures $S(H, E)$ and $N(H, E)$.

4 Rule Induction with BRACID

BRACID is a specialized algorithm to learn rules from imbalanced data. For its details see [16]. Here, we summarize its main characteristics:

- **Hybrid representation of rules and instances**: BRACID tries to create a general description in regions where the examples form large disjuncts (using rules) and instances to better approximate the more difficult decision boundaries. BRACID allows some (difficult) examples to remain not generalized to rules. They can be treated as maximally specific rules.

---

[1] $c_3(H, E) = A(H, E)Z(H, E)$ in case of confirmation and
$c_3(H, E) = -A(H, E)Z(H, E)$ in case of disconfirmation
where
$Z(H, E) = 1 - P(\neg H|E) \div P(\neg H)$ in case of confirmation and
$Z(H, E) = P(H|E) \div P(H) - 1$ in case of disconfirmation;
$A(H, E) = [P(E|H) - P(E)] \div [1 - P(E)]$ in case of confirmation and
$A(H, E) = [P(H) - P(H|\neg E)] \div [1 - P(H)]$ in case of disconfirmation.

- **Bottom-up rule induction**: Unlike a top-down strategy typical for rule induction, BRACID follows a bottom-up (or a specific-to-general) strategy as a more appropriate for imbalanced data. It starts from the set of most specific rules each covering a single learning example – which is called a seed of the rule. Then, in every iteration each rule is generalized in the direction of the nearest neighbour example from the same class, provided that it does not decrease the classification abilities of the whole rule set. The procedure is repeated until no rule in the set can be further generalized.
- **Resignation from greedy, sequential covering technique**: As this technique, popular in typical rule learning algorithms, increases the data fragmentation and is problematic for the minority examples, BRACID takes into account all the learning examples when evaluating new rule candidate.
- **Facing borderline minority examples**: Types of learning examples are evaluated and rules are generated differently depending on the type of the seed example of a rule [17]. The minority examples belonging to the borderline region are allowed to be generalized into more than one rule, to lessen the dominance of the majority class in this region.
- **Facing noisy examples from the majority class**: Noisy majority examples, present inside the minority class regions, may hinder the induction of general minority rules. BRACID has an embedded mechanism for detecting and removing such examples from the learning data set.
- **Less biased classification strategy**: BRACID employs a classification strategy based on nearest rules to diminish the domination of strong majority rules during solving conflict situations while a new instance matches condition parts of many rules.

Note that some mechanisms employed in this algorithm lead to the increase of the number of rules (mainly a bottom-up rule induction and generation of more rules in the borderline regions). However, the increased number of rules for the minority class, coupled with an increased rule support, are beneficial for final classification. The experimental evaluation of classification performance of BRACID showed indeed that it significantly outperformed many standard rule classifiers (induced by RIPPER, PART, C4.5rules, and others) as well as other rule approaches specialized for class imbalance such as modifications of rule search and classification strategies, or the best standard algorithms (e.g., PART) combined with SMOTE methods transforming class distributions [16].

5   Selecting Rules with Respect to Confirmation

We aim to select a subset of induced rules with respect to appropriate rule evaluation measures. In [18] we have already postulated that it would be profitable to find rules which cover diverse sets of examples referring to different sub-parts of the class distribution. Focusing the expert's attention on a subset of rules having such characteristics should be particularly good for the minority class which is often decomposed into many rare sub-concepts.

Recall that several post-pruning techniques have already been proposed to order rules or to reduce their number. However, as we discussed in [18], it may not lead to diverse subsets of rules in BRACID, as e.g. high supports may characterize many rules having similar syntax and covering similar subsets of learning examples. Other post-pruning techniques considered in rule classifiers are focused on optimizing the predictive performance of the rules rather than on improving their descriptive properties [5].

Therefore, we follow a different inspiration, coming from using rules to represent patterns in *subgroup discovery*, where the task is to find subgroups of individuals that are statistically "most interesting" (e.g. covering as many examples as possible and having the most unusual statistical characteristics [5]). In our opinion these kinds of local, diverse patterns correspond to decomposition of the minority class in sub-concepts. In this paper we generalize the algorithm originally proposed in [6] to find rules describing subgroups.

Our approach to select a given number of diverse rules with respect to a given rule evaluation measure is presented in Algorithm 1. It is run for each class separately and takes as an input the set of all rules induced for this class and their required number after selection – later on we discuss how to tune it.

Algorithm 1. Rule Filtering Algorithm

Input: Set of Rules SR for class P, required NUMBER of rules; rule evaluation ev;
Output: Pruned set of rules FR

  Delete rules with too low confirmation from SR
  $FR \leftarrow \emptyset$
  for every example $e \in P$ **do**
    $c(e) \leftarrow 1$
  end for
  repeat
    for each rule $R \in SR$ **do**
      calculate rule evaluation measure $ev(R)$
    end for
    Select $R_{max} = \arg\max_R(ev(R))$
    for each e covered by R_{max} **do**
      $c(e) \leftarrow c(e) + 1$
    end for
    Remove R_{max} from SR
    $FR = FR \cup R_{max}$
  until size of FR = NUMBER

Firstly, we remove all rules with the non-positive value of a selected confirmation measure (except the option where rules are evaluated with the support only). The key idea of the algorithm is to assign a weight $c(e)$ to each learning example. It is initialized with $c(e) = 1$ for all examples from the given class. When rule R is selected, then weights for examples covered by this rule are increased by adding 1. Then, while evaluating the next rule being a candidate for selection, the example takes part in all calculation with the weight $1/c(e)$.

For instance, the support of a rule is computed as a sum of $1/c(e)$ for all target class examples covered by this rule.

This weighted coverage causes that in the subsequent iterations of the algorithm, examples already covered by the selected rules contribute less to the evaluation of new rule. It promotes the rules referring to examples not yet covered and directs the search toward diverse regions of the class.

In this study we will consider three different versions of the rule evaluation $ev(R)$[2] for selecting rules:

1. a standard rule support $sup(R)$;
2. a product of support with a confirmation measure S: $sup(R) \times S(R)$;
3. a product $sup(R) \times N(R)$.

The choice of rule support $sup(R)$ results from earlier experiments in [18] and we want to consider it as a baseline. The choice of both confirmation measures S and N has been justified in Sect. 3. We want to aggregate them with a rule support to represent a trade off in a bi-criteria evaluation where the user is interested in sufficiently strong patterns describing the classes.

6   Experimental Evaluation

In the experiments we will verify whether the proposed post-pruning strategies select a limited number of BRACID rules having better values of interestingness measures than in case of non-pruned rules.

As the evaluation criteria we choose the average values of confirmation measures S and N, rule support and rule confidence. We consider the last two measures due to their popularity in the previous rule filtering techniques and to their easy interpretation for the users. These criteria represent descriptive properties of single rules with respect to their possible interpretability and they are treated as primary criteria in our study. As a secondary criterion, we also evaluate the predictive ability of the rule set, which will be estimated by G-mean and F-measure, both well suited for cases with imbalanced data sets. We use this criterion to control whether pruning the set of rules does not dramatically deteriorate the performance compared to all rules produced by the BRACID algorithm. The predictive measures are evaluated in a repeated stratified 10-fold cross validation procedure while rule evaluation measures are calculated for a set of rules induced from the complete data set.

We analysed previous experiments from [16] and chose 11 data sets where BRACID generated too many rules compared to other, standard rule induction algorithms. They are characterized by different imbalance ratios (from 3% to 30%), data sizes (from 155 to 1728) and types of attributes (only nominal, only numeric, or mixed). Although the imbalance ratios of some of these data sets

---

[2] For simplicity we will further use a notation of a rule as R instead of (H, E) in symbols of measures.

Table 2. Basic characteristics of data sets

Data set	#Examples	Minority class size	Imbalance ratio [%]	#Attributes (numeric)	Minority class name
balance-scale	625	49	7.84	4(4)	B
breast-cancer	286	85	29.72	9(0)	rec-events
car	1728	69	3.99	6(0)	good
cleveland	303	35	11.55	13(6)	positive
cmc	1473	333	22.61	9(2)	long-term
ecoli	336	35	10.42	7(7)	imU
haberman	306	81	26.47	3(3)	died
hepatitis	155	32	20.65	19(6)	die
solar-flareF	1066	43	4.03	12(0)	F
transfusion	748	178	23.80	4(4)	yes
yeast-ME2	1484	51	3.44	8(8)	ME2

are medium, all these data are also affected by different difficulty factors characterizing the distribution of examples from the minority class. According to experimental studies [17] these factors lead to difficulties while learning rules.

All these data sets come from the UCI repository. We analyzed them as binary problems – the minority class vs. majority one (which may aggregate others), as it is a typical view of class imbalances with focusing attention on improving recognition of the class of special importance. The basic characteristics of these data sets are presented in Table 2.

We checked that for all data sets (except cleveland and hepatitis), the BRACID rule sets contained some rules with negative values of confirmation measures. For instance, balance-scale contained 8, car 36, cmc 19, solar-flareF 18 and transfusion 14 such rules.

While using the algorithm for selecting rules we need to define a number of required rules as the stopping condition. In general, this parameter should represent the analyst's expectations and his abilities to inspect the rules. Here we recall our previous experiments [18], where we studied a wide range of values of this parameter (up to 30%). The results showed that the threshold 10% often led to rule sets having the good average rule support and comparable classification performance as the original set of BRACID rules.

Yet another option is to select all the rules which are necessary to cover all the learning examples in each class. We studied this coverage option in [18] and observed that it usually produced higher classification prediction (with respect to G-mean or sensitivity measure) than the percentage option. However, it also selected more rules than the percentage option. As in this study we aim at reducing the number of rules, we decided to consider the percentage option with the parameter tuned to 10% of the original set of rules for each class[3].

---

[3] More detailed experimental results, including also the coverage option are provided at the page http://www.cs.put.poznan.pl/iszczech/publications/nfmcp-2016.html.

Table 3. Characteristics of filtered rules for the minority class

Data set	Pruning	#Rules	Avg.*sup*	Avg.*conf*	Avg.*S*	Avg.*N*
balance-scale	none	52	2.077	0.611	0.535	0.033
	sup	5	6.000	0.266	0.192	0.056
	*sup * S*	5	2.000	0.875	0.799	0.037
	*sup * N*	5	4.600	0.317	0.243	0.065
breast-cancer	none	77	3.364	0.711	0.420	0.030
	sup	8	9.625	0.711	0.434	0.089
	*sup * S*	8	9.125	0.817	0.541	0.094
	*sup * N*	8	10.125	0.736	0.460	0.095
car	none	54	1.444	0.972	0.933	0.021
	sup	5	5.200	0.700	0.663	0.073
	*sup * S*	5	4.800	0.800	0.763	0.068
	*sup * N*	5	5.200	0.700	0.663	0.073
cleveland	none	97	5.495	0.910	0.811	0.154
	sup	10	8.300	0.864	0.773	0.232
	*sup * S*	10	7.300	0.966	0.873	0.207
	*sup * N*	10	8.600	0.864	0.774	0.240
cmc	none	354	6.588	0.723	0.500	0.016
	sup	35	14.914	0.666	0.447	0.037
	*sup * S*	35	12.686	0.782	0.562	0.033
	*sup * N*	35	18.571	0.652	0.434	0.046
ecoli	none	46	10.413	0.872	0.796	0.291
	sup	5	17.400	0.802	0.746	0.483
	*sup * S*	5	17.000	0.889	0.832	0.478
	*sup * N*	5	18.200	0.788	0.734	0.503
haberman	none	122	6.049	0.716	0.464	0.062
	sup	12	9.917	0.650	0.406	0.099
	*sup * S*	12	9.417	0.900	0.658	0.109
	*sup * N*	12	12.250	0.783	0.546	0.135
hepatitis	none	66	7.424	0.986	0.820	0.231
	sup	7	12.000	0.971	0.832	0.373
	*sup * S*	7	12.571	1.000	0.864	0.393
	*sup * N*	7	12.571	1.000	0.864	0.393
solar-flareF	none	39	3.051	0.527	0.490	0.066
	sup	4	6.750	0.362	0.327	0.142
	*sup * S*	4	4.500	0.790	0.753	0.102
	*sup * N*	4	7.750	0.382	0.348	0.164
transfusion	none	161	6.360	0.673	0.440	0.028
	sup	16	16.062	0.630	0.404	0.067
	*sup * S*	16	15.562	0.768	0.543	0.071
	*sup * N*	16	18.500	0.679	0.456	0.083
yeast-ME2	none	155	7.432	0.905	0.875	0.145
	sup	16	9.375	0.915	0.886	0.183
	*sup * S*	16	8.875	0.944	0.915	0.174
	*sup * N*	16	10.688	0.904	0.877	0.209

Table 4. Characteristics of filtered rules for the majority class

Data set	Pruning	#Rules	Avg.sup	Avg.$conf$	Avg.S	Avg.N
balance-scale	none	306	12.889	0.996	0.076	0.021
	sup	31	30.097	0.994	0.076	0.049
	$sup * S$	31	30.452	0.997	0.079	0.051
	$sup * N$	31	34.194	0.996	0.079	0.057
breast-cancer	none	75	4.973	0.959	0.261	0.022
	sup	8	11.750	0.925	0.234	0.050
	$sup * S$	8	12.500	0.994	0.304	0.061
	$sup * N$	8	13.375	0.994	0.305	0.065
car	none	69	68.478	0.924	−0.036	0.017
	sup	7	361.286	0.987	0.037	0.187
	$sup * S$	7	351.429	1.000	0.051	0.212
	$sup * N$	7	356.571	1.000	0.051	0.215
cleveland	none	94	16.426	1.000	0.123	0.061
	sup	9	53.444	1.000	0.142	0.199
	$sup * S$	9	53.444	1.000	0.142	0.199
	$sup * N$	9	54.111	1.000	0.142	0.202
cmc	none	401	7.302	0.971	0.198	0.006
	sup	40	21.725	0.975	0.204	0.017
	$sup * S$	40	21.500	0.987	0.217	0.018
	$sup * N$	40	22.975	0.986	0.216	0.019
ecoli	none	47	64.128	0.990	0.141	0.210
	sup	5	207.800	0.999	0.271	0.685
	$sup * S$	5	208.000	0.999	0.271	0.685
	$sup * N$	5	208.000	0.999	0.271	0.685
haberman	none	60	6.383	0.977	0.247	0.027
	sup	6	15.833	0.990	0.269	0.068
	$sup * S$	6	15.833	0.990	0.269	0.068
	$sup * N$	6	15.833	0.990	0.269	0.068
hepatitis	none	52	18.615	1.000	0.241	0.151
	sup	5	59.600	1.000	0.341	0.485
	$sup * S$	5	59.600	1.000	0.341	0.485
	$sup * N$	5	65.200	1.000	0.357	0.530
solar-flareF	none	64	27.781	0.957	−0.002	0.012
	sup	6	165.333	0.982	0.031	0.123
	$sup * S$	6	158.500	0.989	0.039	0.128
	$sup * N$	6	163.833	0.986	0.036	0.129
transfusion	none	118	11.720	0.965	0.206	0.016
	sup	12	51.500	0.932	0.183	0.064
	$sup * S$	12	41.750	0.959	0.209	0.060
	$sup * N$	12	51.750	0.947	0.200	0.073
yeast-ME2	none	613	204.979	1.000	0.041	0.143
	sup	61	514.000	1.000	0.055	0.358
	$sup * S$	61	566.131	1.000	0.057	0.395
	$sup * N$	61	609.197	1.000	0.059	0.425

In our study, we will examine three proposed strategies to select rules with the rule evaluation $ev(R)$ (see Sect. 5), defined as: (1) a standard rule support $sup(R)$; (2) a product $sup(R) \times S(R)$; and (3) a product $sup(R) \times N(R)$.

The rule characteristics with respect to considered criteria are given in Tables 3 and 4, for the minority and majority class, respectively. The column "pruning" corresponds to the selection strategy (note that results for using the standard version of BRACID without pruning is presented in the first row for each data set with an abbreviation "none").

Additionally, we constructed rule classifiers with the three filtering strategies and evaluated their classification performance. The values of G-mean and F-measure are presented in Table 5.

Table 5. G-mean and F-measure for BRACID with all rules vs. filtered rules

Data set	G-mean				F-measure			
	BRACID	sup	$sup*S$	$sup*N$	BRACID	sup	$sup*S$	$sup*N$
balance-scale	0.56	0.63	0.59	0.60	0.19	0.23	0.22	0.21
breast-cancer	0.56	0.59	0.61	0.61	0.44	0.48	0.49	0.49
car	0.88	0.60	0.61	0.64	0.73	0.41	0.42	0.44
cleveland	0.57	0.71	0.72	0.73	0.33	0.41	0.41	0.42
cmc	0.64	0.64	0.64	0.64	0.45	0.45	0.45	0.45
ecoli	0.83	0.85	0.85	0.84	0.60	0.55	0.54	0.55
haberman	0.58	0.54	0.54	0.54	0.44	0.44	0.43	0.43
hepatitis	0.75	0.76	0.75	0.74	0.60	0.59	0.57	0.54
solar-flareF	0.64	0.73	0.65	0.73	0.28	0.32	0.32	0.31
transfusion	0.64	0.63	0.63	0.65	0.47	0.47	0.46	0.48
yeast-ME2	0.71	0.72	0.73	0.77	0.42	0.40	0.40	0.38

7 Discussion of the Experiments

Each of the filtering strategies improves the interestingness measure used in the given strategy. Note that all of them improve average rule supports for both minority and majority classes. For some data sets these improvements are quite high, for instance, for cmc data the average rule supports increase from 6.59 to 18.57 examples in the minority class, and from 7.30 to 22.98 examples in the majority class. Similar high improvements also occur for car, solar flare, ecoli and transfusion data.

The third strategy (based on $sup(R)$ and $N(R)$) increases the average value of measure N for all data sets in both classes—see e.g. hepatitis data, where the improvements are from 0.23 to 0.39 for the minority class and from 0.15 to 0.53 for the majority class. Similar increases have been observed for other data. Similarly, the second strategy (based on $sup(R)$ and $S(R)$) improves the average values of the confirmation measure S – however, it is more visible for the minority class than for the majority one, for instance changes from 0.46 to 0.65 in the minority class and from 0.25 to 0.27 in the majority one for haberman data. Note that values of the confirmation measure S are always higher than N.

It is worth observing that the proposed strategies also improve rule evaluation measures other than the ones used in each strategy. In particular, the third strategy usually provides the highest values of the average support – in the majority of data sets it is better than the first strategy that uses the support only. Although it sometimes slightly improves the confirmation measure S, it usually decreases the average confidence of rules. On the other hand, the second strategy offers the highest increases of the rule confidence. It is more visible for the minority class as the confidence of majority rules is already quite high.

What is also interesting, classification performance of such filtered rules does not decrease too much compared to the original set of rules and for few data it is even better – see results in Table 5.

The differences in results obtained by strategies using S and N measures could be explained by analyzing their formulae (see Eqs. 1 and 2). They exploit the contingency matrix in a different, although symmetric, way. Measure S is more focused on considering a pair of numbers (a and c) decreased by (b and d), while N aggregates a different combination. As BRACID tries to induce rules with a very high confidence (which refers to the pair a and c), it is naturally oriented on obtaining higher values of the S measure. On the other hand, as measure N exploits complementary information to the one used in BRACID rule induction process, it may better co-operate with the rule support in the pruning strategy and may lead to better descriptive rule evaluation as well as classification results.

8   Conclusions and Final Remarks

To sum up, our experiments have clearly demonstrated that all proposed filtering strategies lead to selecting a much smaller number of BRACID induced rules, which are characterized by better values of considered interestingness measures than in case of non-pruned rules.

As future research, we plan to extend the experimental evaluation with other rule classifiers specialized for class imbalances in order to show the generality of our approach. We also intend to confront our pruning strategies with a baseline approach involving a simple rule filtering. Furthermore, we plan to investigate a more local way of calculating the interestingness measures, which will be based on the analysis of neighbor examples to the given rule rather than on all data elements as it is currently done.

Acknowledgement. The research was supported by NCN grant DEC-2013/11/B/ ST6/00963.

References

1. Bayardo, R., Agrawal, R.: Mining the most interesting rules. In: Proceedings of 5th ACM SIGKDD Conference on Knowledge Discovery and Data Mining, pp. 145–154 (1999)
2. Christensen, D.: Measuring confirmation. J. Philos. **96**, 437–461 (1999)
3. Fitelson, B.: The plurality of Bayesian measures of confirmation and the problem of measure sensitivity. Philos. Sci. **66**, 362–378 (1999)
4. Freitas, A.: On rule interestingness measures. Knowl.-Based Syst. **12**, 309–315 (1999)
5. Furnkranz, J., Gamberger, D., Lavrac, N.: Foundations of Rule Learning. Springer, Berlin (2012). doi:10.1007/978-3-540-75197-7
6. Gamberger, D., Lavrac, N.: Expert-guided subgroup discovery: methodology and application. J. Artif. Int. Res. **17**(1), 501–527 (2002)
7. Geng, L., Hamilton, H.: Interestingness measures for data mining: a survey. ACM Comput. Surv. **38**(3), 9 (2006)
8. Glass, D.: Confirmation measures of association rule interestingness. Knowl.-Based Syst. **44**, 65–77 (2013)
9. Greco, S., Slowinski, R., Szczech, I.: Properties of rule interestingness measures and alternative approaches to normalization of measures. Inf. Sci. **216**, 1–16 (2012)
10. Greco, S., Slowinski, R., Szczech, I.: Measures of rule interestingness in various perspectives of confirmation. Inf. Sci. **346**, 216–235 (2016)
11. He, H., Yungian, M. (eds.): Imbalanced Learning. Foundations, Algorithms and Applications. IEEE - Wiley, Hoboken (2013)
12. Heravi, M., Zaiane, O.R.: A study on interestingness measures for associative classifiers. In: Proceedings of ACM-SAC 2010 Conference Track on Data Mining, pp. 1040–1047 (2010)
13. Lavrač, N., Flach, P., Zupan, B.: Rule evaluation measures: a unifying view. In: Džeroski, S., Flach, P. (eds.) ILP 1999. LNCS (LNAI), vol. 1634, pp. 174–185. Springer, Heidelberg (1999). doi:10.1007/3-540-48751-4_17
14. Lenca, P., Vaillant, B., Meyer, P., Lallich, S.: Associations rule interestingness measures: experimental and theoretical studies. In: Guillet, F., Hamilton, H.J. (eds.) Quality Measures in Data Mining. SCI, vol. 43, pp. 51–76. Springer, Heidelberg (2007). doi:10.1007/978-3-540-44918-8_3
15. McGarry, K.: A survey of interestingness measures for knowledge discovery. Knowl. Eng. Rev. **20**(1), 39–61 (2005)
16. Napierala, K., Stefanowski, J.: BRACID: a comprehensive approach to learning rules from imbalanced data. J. Intell. Inf. Syst. **39**(2), 335–373 (2012)
17. Napierala, K., Stefanowski, J.: Types of minority class examples and their influence on learning classifiers from imbalanced data. J. Intell. Inf. Syst. **46**(3), 563–597 (2016)
18. Napierala, K., Stefanowski, J.: Post-processing of BRACID rules induced from imbalanced data. Fundam. Inform. **148**(1–2), 51–64 (2016)
19. Nozick, R.: Philosophical Explanations. Clarendon Press, Oxford (1981)
20. Stefanowski, J., Vanderpooten, D.: Induction of decision rules in classification and discovery-oriented perspectives. Int. J. Intell. Syst. **16**(1), 13–28 (2001)

Analyzing Time-Decay Effects of Mediating-Objects in Creating Trust-Links

Hiroki Takahashi and Masahiro Kimura[(✉)]

Department of Electronics and Informatics,
Ryukoku University, Otsu 520-2194, Japan
kimura@rins.ryukoku.ac.jp

Abstract. We address the problem of modeling trust network evolution through social communications among users in a social media site. In particular, we focus on a *social trust-link* created between two users having *mediating-objects* such as *mediating-users* and *mediating-items*, and analyze the time-decay effects of mediating-objects on social trust-link creation. To this end, we first introduce the *basic TCM model* that can be regarded as a conventional link prediction method based on mediating-objects, and propose the *TCM model with time-decay* by incorporating an appropriate time-decay function into it. We present an efficient learning method of the proposed model, and apply it to an analysis of social trust-link creation for two real item-review sites. We show that the proposed model significantly outperforms the basic TCM model in terms of prediction performance, and clarify several properties of user behavior for social trust-link creation.

Keywords: Trust-network evolution model · Mediating-user · Mediating-item · Time-decay effect

1 Introduction

The advancement of Social Media such as eBay, Epinions and Facebook has allowed us to construct large trust networks, where a trust-link (u, v) from a user u to a user v indicates that u trusts v and tends to be influenced by v. Previous studies [3,6–8,11,13,14,17,18] have already established the importance of trust in social networks for various processes such as information spreading or search.

Modeling human communication behavior in an online world is an underpinning of mining complex social networks, and a central problem in social network analysis. A trust-link established through *mediating-objects* in a social media site can be regarded as the one created through social communications among users. Representative examples of a mediating-object α from a user u to a user v are as follows. The first one is a *mediating-user* α who has both trust-links from user u and to user v. In this case, user u can meet user v by tracing the trust-links via user α. The second one is a *mediating-item* α for which user u found an activity of user v and thereby knew user v. In the case of a product review

© Springer International Publishing AG 2017
A. Appice et al. (Eds.): NFMCP 2016, LNAI 10312, pp. 99–114, 2017.
DOI: 10.1007/978-3-319-61461-8_7

site, this indicates that user v posted a review for product α, and user u read it with interest. In this paper, we refer a trust-link created between two users having a mediating-object to as a *social trust-link*, and address the problem of modeling the mechanism of how a social trust-link is created in the context of trust network evolution in discrete-time steps.

To confirm the effects of mediating-objects on social trust-link creation, we first introduce a basic model of creating a trust-link under mediating-objects, which is referred to as the *basic TCM model*. Note that the basic TCM model does not reflect when the corresponding mediating-objects were formed, but manages all of them equally. Analyzing to whom a user creates a social trust-link from the viewpoint of mediating-objects is connected with conventional methods for link prediction. In fact, when two representative methods are employed to assess the *mediating-value* of each mediating-object (i.e., its essential value on trust-link creation), the basic TCM model leads to two widely-used methods in the context of link prediction. One is the *naive method* in which all mediating-objects are equally valued regardless of their intrinsic properties. Then, the basic TCM model corresponds to the *Common Neighbor/Feature method* [2,12] for link prediction. The other is the *A-A method* in which the intrinsic properties of mediating-objects are taken into account and less active mediating-objects (e.g., mediating-users that have a smaller number of trust-links, mediating-items for which a smaller number of users perform activities, etc.) are more highly valued. In this case, the basic TCM model corresponds to the *Adamic-Adar method* [2,12] for link prediction. As for creation of a social trust-link, it would in general be reasonable to suppose that older mediating-objects are less influential and recent ones are more influential. However, little attention has been given to analyzing the time-decay effects of mediating-objects in creating a trust-link so far.

To analyze the effects of mediating-objects on social trust-link creation in terms of time-decay, we propose the *TCM model with time-decay* by extending the basic TCM model. As time-decay functions that should be incorporated into the proposed TCM model with time-decay, we adopt two typical ones, *exponential decay* and *power-law decay*. We present an efficient learning method of the proposed model, and apply it to an analysis of social trust-link creation for two real item-review sites. We show that the proposed model significantly outperforms the basic TCM model in terms of prediction performance, the power-law decay can be more suitable than the exponential decay for the time-decay function, and the A-A method can be more effective than the naive method for determining mediating-values. Moreover, by employing the TCM model with power-law decay under the A-A method, we analyze how an individual user creates social trust-links from the perspectives of mediating-user, mediating-item and time-decay, and clarify several properties of user behavior.

2   Related Work

Social trust-link creation based on mediating-users is closely related to the *triadic closure mechanism* which is derived from the concept that two people with mutual friends have a higher chance to create a link, and which was regarded as a powerful principle to explain link creation in online social networks [10]. By reflecting other social theories as well, this mechanism was extended to predict positive (trust) and negative (distrust) links in a signed social network [11]. A method of predicting negative links based on positive links and content-centric interactions was also proposed [16]. Weng et al. [19] extended the triadic closure mechanism by exploring the role of information diffusion in the evolution of a social network, and showed that shortcuts based on information flow are another key factor in explaining link formation.

Modeling network evolution can be connected with link prediction and rating prediction in recommender systems. Various link prediction methods were presented in this context, including supervised trust prediction [13,14], non-negative matrix factorization based on both link and rating information [17], link prediction with explanations for user recommendation systems [2], link prediction from information diffusion data [5], and link prediction in multiple networks [20]. To represent the temporal change of user preference in a recommender system, Koren [9] incorporated a time decay function, and improved the performance in rating prediction. Tang et al. [18] enhanced this framework by combining trust network information, and demonstrated its effectiveness in rating prediction and link prediction. Note that these researches were not intended for directly modeling the dynamics of trust network evolution.

Unlike the previous work such as the approaches mentioned above, we focus on modeling the mechanism of creating trust-links under the presence of mediating-objects, and deal with both mediating-user and mediating-item as mediating-object. Furthermore, we provide a novel model and its efficient learning method in order to analyze the time-decay effects of mediating-objects and the difference of mediating-user and mediating-item in influence strength.

3   Analysis Model

3.1   Problem Formulation

For a social media site offering trust-links and activities, we investigate the evolution of trust network in a given time-period $J = [T_f, T_\ell)$ in discrete-time steps, where T_f and T_ℓ are positive integers with $T_f < T_\ell$. Here, we assume that J is not so long (e.g., around six months) since users' behavioral patterns and preferences in an online world can largely change over a long time frame in general. We focus on a social trust-link (i.e., a trust-link created between two users having a mediating-object), and consider analyzing the effects of mediating-objects in creating social trust-links. For each mediating-object α from a user u to a user v, let $\tau_\alpha(u, v)$ denote the time-step at which α first became a mediating-object from u to v. Here, $\tau_\alpha(u, v)$ is called the *mediating-time* of α from u to v. Since validity period of information cannot in general be so long in an online world, we only treat such trust-links and activities that have been relatively recently generated. Thus, we regard α as a mediating-object from u to v at a time-step t if and only if $t - \Delta t_0 \leq \tau_\alpha(u, v) < t$, where Δt_0 is a positive integer (e.g., around three months) specified in advance, and stands for the validity period of information in this social media site. We also assume that the set of mediating-objects are divided into K (≥ 2) kinds of categories including "Mediating-user", "Mediating-item", etc.

Let U be the set of all users in the site during time-period J. For an arbitrary time-step $t \in J$, let U_t denote the set of all users at time-step t, and let $\bar{E}_t$ ($\subset U \times U$) denote the set of all trust-links created in the set of users U before time-step t. Then, we have $U = \bigcup_{t \in J} U_t$, and $\bar{E}_{t+1} \setminus \bar{E}_t$ indicates the set of all trust-links created in U at time-step t. For any $u, v \in U_t$ and $k \in \{1, \ldots, K\}$, let $\mathcal{M}_{k,t}(u, v)$ denote the set of all mediating-objects of category k from user u to user v at time step t. We assume that $\mathcal{M}_{k,t}(u, v) \cap \mathcal{M}_{\ell,t}(u, v) = \emptyset$ if $k \neq \ell$. For any $u \in U_t$, we define $V_t(u)$ by

$$V_t(u) = \left\{ v \in U_t \ \middle| \ \bigcup_{k=1}^{K} \mathcal{M}_{k,t}(u, v) \neq \emptyset, \ (u, v) \notin \bar{E}_t \right\}.$$

Note that a social trust-link created at a time-step $t \in J$ is represented by (u, v), where $u \in U_t$ and $v \in V_t(u)$.

Suppose that a user $u \in U_t$ creates a social trust-link to some user belonging to $V_t(u)$ at a time-step $t \in J$. Then, to analyze the effects of mediating-objects on social trust-link creation, we consider modeling the probability $P(v \mid u, t)$ that the user u creates a trust-link (u, v) to a user $v \in V_t(u)$ at the time-step t. It is conceivable that the influence of a mediating-object $\alpha \in \mathcal{M}_{k,t}(u, v)$ depends on how close the mediating-time $\tau_\alpha(u, v)$ is to the time-step t. Moreover, it is expected that the larger $t - \tau_\alpha(u, v)$ is, the smaller the influence of α becomes. We can also speculate that how influential a mediating-object $\alpha \in \mathcal{M}_{k,t}(u, v)$ is on the creation of trust-link (u, v) depends on its category k. In this paper, we construct such a model of probability $P(v \mid u, t)$ that can analyze the effects of time-decay and category for mediating-objects in creating social trust-links.

3.2   Basic Model

First, we introduce a basic model for evaluating the effects of mediating-objects on social trust-link creation, which is shown to be associated with a conventional method for link prediction when a widely-used simple method is employed to assess the essential value of a mediating-object on the basis of its observed features (see Sect. 5.2). Here, we fix such an assessment method, and determine the *mediating-value* of each mediating-object $\alpha \in \mathcal{M}_{k,t}(u, v)$ of category k from user u to user v at time-step t. Let $r_{k,t}(\alpha)$ denote the mediating-value of α, where $0 < r_{k,t}(\alpha) \leq 1$.

It can be expected that the probability $P(v \mid u, t)$ becomes high if there are many mediating-objects of high mediating-values from user u to user $v \in V_t(u)$ at time-step t. Thus, as a model of probability $P(v \mid u, t)$ for $v \in V_t(u)$, we define

$$P^{basic}(v \mid u, t) \propto \sum_{k=1}^{K} \sum_{\alpha \in \mathcal{M}_{k,t}(u,v)} r_{k,t}(\alpha) \quad (\forall v \in V_t(u)), \tag{1}$$

where it is set that $\sum_{\alpha \in \mathcal{M}_{k,t}(u,v)} r_{k,t}(\alpha) = 0$ if $\mathcal{M}_{k,t}(u,v) = \emptyset$. This model (see Eq. (1)) can be regarded as a basic model of trust-link creation based on mediating-objects, and is referred to as the *basic TCM model*.

3.3 Proposed Model

In order to analyze the effects of mediating-objects on creation of social trust-links in terms of time-decay and category, we consider extending the basic TCM model, and propose modeling probability $P(v \mid u, t)$ as

$$P^{decay}(v \mid u, t; \boldsymbol{\lambda}, \boldsymbol{\mu}) \propto \sum_{k=1}^{K} e^{\mu_k} \sum_{\alpha \in \mathcal{M}_{k,t}(u,v)} r_{k,t}(\alpha) f(t - \tau_\alpha(u,v); \lambda_k) \quad (\forall v \in V_t(u)),$$

(2)

where $\boldsymbol{\lambda} = (\lambda_1, \ldots, \lambda_K)$, $(\lambda_1, \ldots, \lambda_K > 0)$ and $\boldsymbol{\mu} = (\mu_1, \ldots, \mu_K)$, $(\mu_1, \ldots, \mu_K \in \mathbb{R})$ are the model parameters whose values are estimated from the observed data, and it is set that $\sum_{\alpha \in \mathcal{M}_{k,t}(u,v)} r_{k,t}(\alpha) f(t - \tau_\alpha(u,v); \lambda_k) = 0$ if $\mathcal{M}_{k,t}(u,v) = \emptyset$. Here, for each $k \in \{1, \ldots, K\}$, λ_k represents the time-decay rate of a mediating-object belonging to category k, and e^{μ_k} represents the relative influence strength of mediating-objects belonging to category k. Moreover, $f(s; \lambda_k)$ is a monotone decreasing function for $s > 0$, and models a time-decay effect. In this paper, we adopt two typical time-decay functions related to human behavior [1,9,15]. One is

$$f_{ex}(s; \lambda_k) = C_0 e^{-\lambda_k s}$$

(3)

for $s > 0$, which is called the *exponential decay function*, and the other is

$$f_{pl}(s; \lambda_k) = C_1 s^{-\lambda_k}$$

(4)

for $s > 0$, which is called the *power-law decay function*. Here, C_0 and C_1 are the normalization constants with $C_0, C_1 > 0$.

The proposed analysis model (see Eq. (2)) is referred to as the *TCM model with time-decay*. In particular, the TCM model with time-decay function $f_{ex}(s; \lambda_k)$ and the TCM model with time-decay function $f_{pl}(s; \lambda_k)$ are called the *TCM model with exponential decay* and the *TCM model with power-law decay*, respectively.

4 Parameter Estimation Method

Let $D_* = \{(u, v, t)\}$ denote the set of all social trust-links created within a time-period $J_* = [T_f, T_*)$, where T_* is a positive integer with $T_f < T_* \le T_\ell$. Here, $(u, v, t) \in D_*$ indicates that user u created a social trust-link to user v at time-step $t \in J_*$. Then, we consider estimating the parameter values of the TCM model with time-decay from the observed data D_*. Note that to determine the mediating-time $\tau_\alpha(u, v)$ of a mediating-object α from a user u to a user v in the observed time-period J_*, the data in the time-period $J' = [T_f - \Delta t_0, T_f)$ before J_* is also required for the parameter estimation.

To estimate the values of $\boldsymbol{\lambda} = (\lambda_1, \ldots, \lambda_K)$ and $\boldsymbol{\mu} = (\mu_1, \ldots, \mu_K)$ from D_*, we conform to the framework of MAP estimation, and consider maximizing the function

$$\mathcal{L}(\boldsymbol{\lambda}, \boldsymbol{\mu}) = \sum_{(u,v,t) \in D_*} \log P^{decay}(v \mid u, t; \boldsymbol{\lambda}, \boldsymbol{\mu}) + \sum_{k=1}^{K} \left((b_k - 1) \log \lambda_k - c_k \lambda_k - \frac{\mu_k{}^2}{2d_k{}^2} \right) \tag{5}$$

with respect to $\boldsymbol{\lambda}$ and $\boldsymbol{\mu}$ (see Eqs. (2), (3) and (4)), where $b_k \geq 1$, $c_k > 0$ and $d_k > 0$ are regularization constants for $k = 1, \ldots, K$. Here, we assume a gamma prior for each λ_k and a Gaussian prior for each μ_k. We consider deriving an iterative algorithm. Let $\bar{\boldsymbol{\lambda}}$ and $\bar{\boldsymbol{\mu}}$ denote the current estimates of $\boldsymbol{\lambda}$ and $\boldsymbol{\mu}$, respectively. By applying Jensen's inequality, we can obtain a convex function $Q(\boldsymbol{\lambda}, \boldsymbol{\mu} \mid \bar{\boldsymbol{\lambda}}, \bar{\boldsymbol{\mu}})$ of $\boldsymbol{\lambda}$ and $\boldsymbol{\mu}$ such that $\mathcal{L}(\boldsymbol{\lambda}, \boldsymbol{\mu}) - \mathcal{L}(\bar{\boldsymbol{\lambda}}, \bar{\boldsymbol{\mu}}) \geq Q(\boldsymbol{\lambda}, \boldsymbol{\mu} \mid \bar{\boldsymbol{\lambda}}, \bar{\boldsymbol{\mu}})$ and $Q(\bar{\boldsymbol{\lambda}}, \bar{\boldsymbol{\mu}} \mid \bar{\boldsymbol{\lambda}}, \bar{\boldsymbol{\mu}}) = 0$. Thus, we can derive an update formula for $\boldsymbol{\lambda}$ and $\boldsymbol{\mu}$ by maximizing $Q(\boldsymbol{\lambda}, \boldsymbol{\mu} \mid \bar{\boldsymbol{\lambda}}, \bar{\boldsymbol{\mu}})$ (see Appendix for the details of the learning algorithm).

5　Experiments

5.1　Social Media Data

We collected real data from two social media sites, Epinions[1] and @cosme[2], where Epinions is a product-review site, and @cosme is a Japanese word-of-mouth communication site for cosmetics. In both sites, a user can create a trust-link to another user, and post a review for an item. As for Epinions, we traced the trust-links by the breadth-first search from a user who was featured as the most popular user in October 2012 until no new users appeared, and collected sets of trust-links and reviews. In a similar way, we also collected such data for @cosme in June 2010. We aggregated the data into day granularity (i.e., one time-step is set as one day). The collected data includes 64,268 users, 509,293 trust-links and 809,517 reviews for 268,891 items for Epinions, and 30,369 users, 359,817 trust-links, 3,815,622 reviews for 122,927 items for @cosme.

We confirmed that all of the indegree distribution (i.e., the fraction of the number of trust-links a user received), the outdegree distribution (i.e., the fraction of the number of trust-links a user created) and the activity distribution (i.e., the fraction of the number of reviews a user posted) exhibit power-law tails, which are known as typical properties of social data in an online world. By taking into consideration the fact that trust-links were constantly generated in 2003, we constructed datasets from the trust-links and the reviews generated in October 2002 to December 2003 for both sites. These datasets are referred to as the Epinions data and the @cosme data, respectively.

---

[1] http://www.epinions.com/.
[2] http://www.cosme.net/.

5.2   Definition of Mediating-Objects

As for mediating-objects to be examined, we focused on two categories (i.e., $K = 2$); category 1 is "Mediating-user" and category 2 is "Mediating-item". For each $t \in J$, let E_t ($\subset U \times U$) denote the set of all trust-links created in the set of users U within time-period $[t - \Delta t_0, t)$. For each $t \in J$ and $u \in U$, let $A_t(u)$ denote the set of items for which user u posted reviews within time-period $[t - \Delta t_0, t)$. Here, by considering the volume of data involved, we simply set the validity period Δt_0 as three months. For each $(u, w) \in E_t$ and $a \in A_t(u)$, let $T_1(u, w)$ denote the time-step at which user u creates trust-link (u, w), and let $T_2(u, a)$ denote the time-step at which user u posts a review for item a. First, we define the category "Mediating-user" as follows: A user α is a *mediating-user* from a user u to a user v at a time-step t when there exist trust-links $(u, \alpha), (\alpha, v) \in E_t$. For a mediating-user α from user u to user v at time-step t, we have $t - \Delta t_0 \leq T_1(u, \alpha), T_1(\alpha, v) < t$, and define the mediating-time $\tau_\alpha(u, v)$ as the maximum of $T_1(u, \alpha)$ and $T_1(\alpha, v)$. Next, we define the category "Mediating-item" as follows: An item α is a *mediating-item* from a user u to a user v at a time-step t when (1) $\alpha \in A_t(u) \cap A_t(v)$ or (2) there exists some user $w \in U_t$ such that $(u, w) \in E_t$ and $\alpha \in A_t(w) \cap A_t(v)$. For a mediating-item α from user u to user v at time-step t, we have $t - \Delta t_0 \leq T_2(w, \alpha), T_2(\alpha, v) < t$, and also define the mediating-time $\tau_\alpha(u, v)$ as the maximum of $T_2(u, \alpha)$ and $T_2(\alpha, v)$. Here, we note that to identify the items for which user v has recently posted reviews and user u can read those reviews with interest, we examine not only the items for which user u has recently posted reviews but also the items for which the users to whom user u has recently created trust-links has recently posted reviews.

For the Epinions and @cosme data, we investigated a relationship between the creation of a social trust-link and the number of mediating-objects according to the work of Crandall et al. [4]. In 2003, there were 7,965 and 8,699 social trust-links for the Epinions data and the @cosme data, respectively. For such a social trust-link (u, v), we examined change in the number of mediating-objects from user u to user v as a function of the number of days after the social trust-link (u, v) was created. Figure 1 indicates change in the average number of mediating users and items for all the social trust-links for the Epinions and @cosme data. We can observe a sharp increase in the numbers of mediating users and items immediately before the social trust-link creation. These results imply that there exists a correlation between the creation of a social trust-link and the number of mediating-objects, and suggest that incorporating the number of mediating-objects can be a promising approach for modeling social trust-link creation.

5.3   Definition of Mediating-Values

As for determining the mediating-value $r_{k,t}(\alpha)$ of a mediating-object $\alpha \in \mathcal{M}_{k,t}(u, v)$ of a category k from a user $u \in U_t$ to a user $v \in V_t(u)$ at a time-step $t \in J$, we employed two representative methods in the experiments.

First, we examined the method of equally assessing all mediating-objects; i.e.,

$$r_{k,t}(\alpha) = 1$$

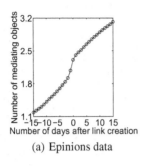
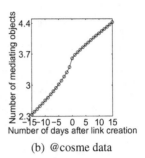

(a) Epinions data   (b) @cosme data

Fig. 1. Relation between the creation of social trust-links and the number of mediating-objects.

for any $\alpha \in \mathcal{M}_{k,t}(u, v)$. This method is referred to as the *naive method* for mediating-values. In this case, we have $P^{base}(v \mid u, t) \propto |\mathcal{M}_{1,t}(u, v)| + |\mathcal{M}_{2,t}(u, v)|$ for $v \in V_t(u)$. Here, note that $\mathcal{M}_{1,t}(u, v) = \mathcal{N}_{1,t}^{out}(u) \cap \mathcal{N}_{1,t}^{in}(v)$ and $\mathcal{M}_{2,t}(u, v) = \left(\bigcup_{w \in \mathcal{N}_{1,t}^{out}(u) \cup \{u\}} A_t(w) \right) \cap A_t(v)$, where $\mathcal{N}_{1,t}^{out}(w) = \{w' \in U_t \mid (w, w') \in E_t\}$ and $\mathcal{N}_{1,t}^{in}(w) = \{w' \in U_t \mid (w', w) \in E_t\}$ for any user $w \in U_t$. Thus, the basic TCM model can be regarded as a kind of Common Neighbor/Feature method [2,12] for link prediction.

Next, we examined a method in which (1) mediating-users having a smaller number of trust-links are more highly valued and (2) mediating-items for which a smaller number of users posted reviews are more highly valued. We defined $r_{k,t}(\alpha)$ by

$$r_{k,t}(\alpha) = \frac{C_2}{|\mathcal{N}_{k,t}(\alpha)|} + C_3$$

for any $\alpha \in \mathcal{M}_{k,t}(u, v)$, where the constants C_2 (> 0) and C_3 were determined as follows: $r_{k,t}(\alpha) = 1$ if $|\mathcal{N}_{k,t}(\alpha)| = 2$ and $r_{k,t}(\alpha) = 0.2$ if $|\mathcal{N}_{1,t}(\alpha)| = \max\{|\mathcal{N}_{1,t'}(\alpha')|; \ t' \in J_0, \ \alpha' \in U_{t'}\}$ for $k = 1$ and $|\mathcal{N}_{2,t}(\alpha)| = \max\{|\mathcal{N}_{2,t'}(\alpha')|; \ t' \in J_0, \ \alpha' \in A\}$ for $k = 2$. Here, $\mathcal{N}_{1,t}(\alpha) = \mathcal{N}_{1,t}^{out}(\alpha) \cup \mathcal{N}_{1,t}^{in}(\alpha)$, $\mathcal{N}_{2,t}(\alpha) = \{w \in U_t \mid \alpha \in A_t(w)\}$ and A is the set of all items. This method is referred to as the *A-A method* for mediating-values. We note that in this case, the basic TCM model can be regarded as a kind of Adamic-Adar method [2,12] for link prediction (see Eq. (1)).

5.4   Datasets

Using the proposed TCM model with time-decay, we analyze how social trust-links are created in time-period $J = [T_f, T_\ell)$ for the Epinions and @cosme data. Let $D = \{(u, v, t)\}$ denote the set of all social trust-links created within time-period J, where $(u, v, t) \in D$ indicates that user u created a social trust-link to user v at time-step t. To evaluate the prediction performance of the proposed model, we divide time-period J into a training time-period $J_0 = [T_f, T_m)$ and a test time-period $J_1 = [T_m, T_\ell)$, and define a training set D_0 and a test set D_1

by $D_0 = \{(u, v, t) \in D \,|\, t \in J_0\}$ and $D_1 = \{(u, v, t) \in D \,|\, t \in J_1\}$, where T_m is a positive integer with $T_f < T_m < T_\ell$. In the experiments, we set $|J_0| = |J_1| = \Delta t_0$; i.e., $T_\ell = T_f + 2\Delta t_0$ and $T_m = T_f + \Delta t_0$. As mentioned before, we also use the data in time-period $J' = [T_f - \Delta t_0, T_f)$ before time-period $J = [T_f, T_\ell)$ in order to determine mediating-time.

The TCM model with time-decay is learned from training set D_0, and its generalization capability is evaluated using test set D_1. Here, the regularization constants were set as $b_k = 1$, $c_k = 1$, $d_k = 0.1$ for each category k. For each of the Epinions and @cosme data, we constructed three datasets $\mathcal{D}_1$, $\mathcal{D}_2$ and $\mathcal{D}_3$ by setting J_0 as January to March in 2003 for $\mathcal{D}_1$, April to June in 2003 for $\mathcal{D}_2$ and July to September in 2003 for $\mathcal{D}_3$, respectively. Here, for example, as for $\mathcal{D}_1$, J_1 is April to June in 2003 and J' is October to December in 2002.

5.5   Evaluation Results

For each of the three datasets $\mathcal{D}_1$, $\mathcal{D}_2$ and $\mathcal{D}_3$ in the Epinions and @cosme data, we estimated the parameter values of the TCM model with time-decay from training set D_0 (see Sect. 4), and evaluated the prediction capability of the learned model using test set D_1 in terms of *prediction log-likelihood ratio PLR* since it is a probabilistic generative model.[3] Here, *PLR* is defined by

$$PLR = \sum_{(u,v,t) \in D_1} \left(\log \hat{P}(v \,|\, u, t) - \log \frac{1}{|V_t(u)|} \right) \qquad (6)$$

where $\hat{P}(v \,|\, u, t)$ stands for a prediction of probability $P(v \,|\, u, t)$ for $(u, v, t) \in D_1$ by a specified model. Note that $1/|V_t(u)|$ indicates the prediction likelihood of the random guessing for $(u, v, t) \in D_1$. We compared the basic TCM model and the proposed TCM models in terms of *PLR*, and examined which model is suitable. Figures 2 and 3 show the results of the basic TCM model (Bs), the TCM model with exponential decay (ED) and the TCM model with power-law decay (PD) for the Epinions data and the @cosme data, respectively. Here, Figs. 2(a)–(c), 3(a)–(c) indicate the results for the naive method for mediating-values, and Figs. 2(d)–(f), 3(d)–(f) indicate the results for the A-A method for mediating-values. *PLR* measures the relative performance versus the random guessing. We observe that the basic TCM model provided much better performance than the random guessing, and reconfirm the importance of exploiting mediating-objects (i.e., the conventional methods for link prediction). Moreover, we see that the TCM model with power-law decay performed the best, the TCM model with exponential decay followed the next, and the basic TCM model was always much worse than these two models. These results demonstrate the effectiveness of incorporating time-decay, and also coincide with the observations that many human behaviors follow power-laws [1,15]. Thus, we focus on the TCM model

---

[3] We also evaluated its prediction capability in terms of the area under the ROC curve (AUC) for trust-link prediction, and confirmed that the results for AUC were similar to those for *PLR*.

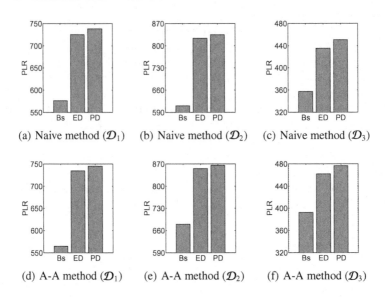

(a) Naive method ($\mathcal{D}_1$)   (b) Naive method ($\mathcal{D}_2$)   (c) Naive method ($\mathcal{D}_3$)

(d) A-A method ($\mathcal{D}_1$)   (e) A-A method ($\mathcal{D}_2$)   (f) A-A method ($\mathcal{D}_3$)

Fig. 2. Evaluation results of the proposed models for the Epinions data.

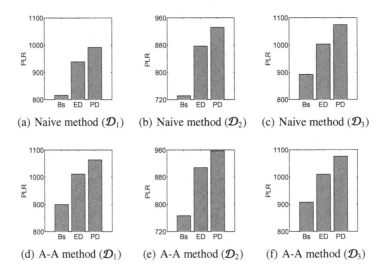

(a) Naive method ($\mathcal{D}_1$)   (b) Naive method ($\mathcal{D}_2$)   (c) Naive method ($\mathcal{D}_3$)

(d) A-A method ($\mathcal{D}_1$)   (e) A-A method ($\mathcal{D}_2$)   (f) A-A method ($\mathcal{D}_3$)

Fig. 3. Evaluation results of the proposed models for the @come data.

with power-law decay. Then, in the matter of determining mediating-values, we see that the A-A method can be more effective than the naive method. Hence, for the behavior analysis, we employed the TCM model with power-law decay under the A-A method for mediating-values.

Table 1 indicates the estimated results for the parameters in the TCM model with power-law decay under the A-A method. We observe that λ_2 was larger

Table 1. The estimated results for the parameters in the TCM model with power-law decay.

	Epinions data					@cosme data			
	λ_1	λ_2	e^{μ_1}	e^{μ_2}		λ_1	λ_2	e^{μ_1}	e^{μ_2}
$\mathcal{D}_1$	0.17	0.47	1.25	0.80	$\mathcal{D}_1$	0.06	0.53	1.96	0.51
$\mathcal{D}_2$	0.17	0.46	1.15	0.87	$\mathcal{D}_2$	0.04	0.42	2.22	0.45
$\mathcal{D}_3$	0.25	0.50	1.05	0.96	$\mathcal{D}_3$	0.06	0.42	1.87	0.54

than λ_1, and e^{μ_1} was larger than e^{μ_2}. These results show that from a system-wide point of view, the time-decay rate of a mediating-item is higher than that of a mediating-user, and mediating-users are more influential than mediating-items. In particular, this implies that the influence of mediating-items tends to decrease more rapidly than that of mediating-users as time passes.

5.6 Analysis Results

Now, we focus on the behavior of an individual user in creating social trust-links, and analyze it by the TCM model with power-law decay under the A-A method. In the experiments, we investigated the users who created at least 5 social trust-links during the entire time-period J. Such a user is referred to as the *analysis target user*. The number of analysis target users was 96 and 46 in dataset $\mathcal{D}_1$, 78 and 59 in dataset $\mathcal{D}_2$, and 87 and 45 in dataset $\mathcal{D}_3$ for the Epinions data and the @cosme data, respectively. For each analysis target user u of each dataset, we estimated the values of parameters λ_1, λ_2, e^{μ_1} and e^{μ_2} in the TCM model with power-law decay by using the data of both the social trust-links created by the user u during the entire time-period J and the corresponding mediating-objects.

Figure 4 shows the analysis results, where it plots e^{μ_2}/e^{μ_1} versus λ_2/λ_1 for all the analysis target users. We observe that the points $(\lambda_2/\lambda_1, e^{\mu_2}/e^{\mu_1})$ plotted on the coordinate plane vary substantially depending on the users. However, the entire tendencies of the results do not depend largely on the datasets. Most users have the property that λ_2 is larger than λ_1, and e^{μ_2} is smaller than e^{μ_1}, that is, the time-decay rate of a mediating-item tends to be higher than that of a mediating-user, and mediating-items tend to be less influential than mediating-users, which coincides with the system-wide property observed in Sect. 5.5. In particular, we observe that @cosme tends to have smaller e^{μ_2}/e^{μ_1} than Epinions. This implies that mediating-users tend to be much more influential than mediating-items in @cosme. Moreover, through our analysis, we can identify several characteristic users, including the users for whom mediating-items are substantially more influential than mediating-users (i.e., $e^{\mu_2} \gg e^{\mu_1}$), and the users for whom the time-decay rate of a mediating-item is substantially lower than that of a mediating-user (i.e., $\lambda_2 \ll \lambda_1$). These results demonstrate the effectiveness of the analysis method based on the TCM model with time-decay.

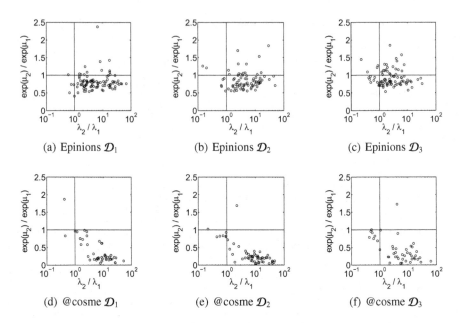

Fig. 4. Analysis results of user behavior by the TCM model with power-law decay.

6 Conclusion

Aiming to construct a better model for trust network evolution through social communications among users in a social media site, we have proposed a novel model that can analyze the effects of time-decay and category with respect to mediating-objects in creating social trust-links. As a basic model of creating social trust-links under mediating-objects, we first introduced the basic TCM model that can be regarded as a conventional link prediction method. In fact, it leads to the Common Neighbor/Feature and the Adamic-Adar methods when the naive and A-A methods are employed to determine mediating-values, respectively. To analyze the effects of mediating-objects in terms of time-decay and category, we proposed the TCM model with time-decay, which derives two models depending on the time-decay functions incorporated; i.e., the TCM model with exponential decay and the TCM model with power-law decay. We presented an efficient method of estimating the values of parameters in the proposed model from the observed data of social trust-links and corresponding mediating-objects. We applied the proposed TCM model with time-decay to real data from two item-review sites "Epinions" and "@cosme", where mediating-users and mediating-items were examined as mediating-objects.

Then, we first showed that the TCM model with time-decay exhibits higher prediction capability than the basic TCM model, demonstrating the effectiveness of incorporating time-decay. We also showed that the TCM model with power-law decay outperforms the TCM model with exponential decay, and as for determining mediating-values, the A-A method is more effective than the naive method. Thus, using the TCM model with power-law decay under the A-A method, we analyzed the behavior of creating social trust-links in the Epinions and @cosme data. From a system-wide point of view, the time-decay rate of a mediating-item was higher than that of a mediating-user, and mediating-users were more influential than mediating-items. Moreover, we identified several characteristic users according to our analysis method. These results demonstrate the effectiveness of the proposed analysis model. Our immediate future work includes extensively evaluating the model for various social media data, investigating various kinds of mediating-objects, and exploring better ways to determine mediating-values.

Acknowledgements. This work was partly supported by JSPS Grant-in-Aid for Scientific Research (C) (No. 26330352), Japan.

Appendix: Learning Algorithm

We consider learning the TCM model with time-decay from the observed data D_*. We derive an iterative algorithm for estimating the values of $\boldsymbol{\lambda}$ and $\boldsymbol{\mu}$ by maximizing the objective function (see Eq. (5)). Let $\bar{\boldsymbol{\lambda}} = (\bar{\lambda}_1, \ldots, \bar{\lambda}_K)$ and $\bar{\boldsymbol{\mu}} = (\bar{\mu}_1, \ldots, \bar{\mu}_K)$ be the current estimates of $\boldsymbol{\lambda}$ and $\boldsymbol{\mu}$, respectively. By Jensen's inequality, we have

$$
\log \left(\sum_{k=1}^{K} e^{\mu_k} \sum_{\alpha \in \mathcal{M}_{k,t}(u,v)} r_{k,t}(\alpha) f(t - \tau_\alpha(u,v); \lambda_k) \right)
$$

$$
- \log \left(\sum_{k=1}^{K} e^{\bar{\mu}_k} \sum_{\alpha \in \mathcal{M}_{k,t}(u,v)} r_{k,t}(\alpha) f(t - \tau_\alpha(u,v); \bar{\lambda}_k) \right)
$$

$$
\geq \sum_{k=1}^{K} \sum_{\alpha \in \mathcal{M}_{k,t}(u,v)} \bar{q}_{k,\alpha}(u,v,t) \log \left(\frac{e^{\mu_k} f(t - \tau_\alpha(u,v); \lambda_k)}{e^{\bar{\mu}_k} f(t - \tau_\alpha(u,v); \bar{\lambda}_k)} \right), \tag{7}
$$

for any $(u,v,t) \in D_*$, where

$$
\bar{q}_{k,\alpha}(u,v,t) = \frac{e^{\bar{\mu}_k} r_{k,t}(\alpha) f(t - \tau_\alpha(u,v); \bar{\lambda}_k)}{\sum_{\ell=1}^{K} \sum_{\beta \in \mathcal{M}_{\ell,t}(u,v)} e^{\bar{\mu}_\ell} r_{\ell,t}(\beta) f(t - \tau_\beta(u,v); \bar{\lambda}_\ell)} > 0 \tag{8}
$$

for $k = 1, \ldots, K$ and $\alpha \in \mathcal{M}_{k,t}(u,v)$. Note that $\sum_{k=1}^{K} \sum_{\alpha \in \mathcal{M}_{u,v,t}^{k}} \bar{q}_{k,\alpha}(u,v,t) = 1$. Thus, by Eqs. (2), (5) and (7), we have $\mathcal{L}(\boldsymbol{\lambda}, \boldsymbol{\mu}) - \mathcal{L}(\bar{\boldsymbol{\lambda}}, \bar{\boldsymbol{\mu}}) \geq Q(\boldsymbol{\lambda}, \boldsymbol{\mu} \,|\, \bar{\boldsymbol{\lambda}}, \bar{\boldsymbol{\mu}})$, where

$$
Q(\boldsymbol{\lambda}, \boldsymbol{\mu} \,|\, \bar{\boldsymbol{\lambda}}, \bar{\boldsymbol{\mu}}) = - \sum_{(u,v,t) \in D_*} \sum_{k=1}^{K} \sum_{\alpha \in \mathcal{M}_{k,t}(u,v)} \bar{q}_{k,\alpha}(u,v,t) \left(g_\alpha(u,v,t) \lambda_k - \mu_k \right)
$$

$$- \sum_{(u,v,t)\in D_*} \log \left(\sum_{w\in V_t(u)} \sum_{k=1}^{K} \sum_{\alpha\in\mathcal{M}_{k,t}(u,w)} \exp(-g_\alpha(u,w,t)\,\lambda_k + \mu_k) \right)$$

$$+ \sum_{k=1}^{K} \left((b_k - 1)\log\lambda_k - c_k\lambda_k - \frac{\mu_k^2}{2d_k^2} \right) + const. \tag{9}$$

Here, for $(u,v,t) \in D_*$, $w \in V_t(u) \cup \{v\}$, $k = 1,\ldots,K$ and $\alpha \in \mathcal{M}_{k,t}(u,w)$, $g_\alpha(u,w,t)$ is defined as follows: $g_\alpha(u,w,t) = t - \tau_\alpha(u,w)$ if $f(s;\lambda_k) = f_{ex}(s;\lambda_k)$ and $g_\alpha(u,w,t) = \log(t - \tau_\alpha(u,w))$ if $f(s;\lambda_k) = f_{pl}(s;\lambda_k)$ (see Eqs. (3) and (4)). Also, $const$ indicates such a constant term that does not depend on $\boldsymbol{\lambda}$ and $\boldsymbol{\mu}$. Note that $Q(\bar{\boldsymbol{\lambda}}, \bar{\boldsymbol{\mu}} \mid \bar{\boldsymbol{\lambda}}, \bar{\boldsymbol{\mu}}) = 0$. Thus, we consider increasing the value of $\mathcal{L}(\boldsymbol{\lambda}, \boldsymbol{\mu})$ by maximizing $Q(\boldsymbol{\lambda}, \boldsymbol{\mu} \mid \bar{\boldsymbol{\lambda}}, \bar{\boldsymbol{\mu}})$. We define $h_{k,\alpha}(u,v,t;\boldsymbol{\lambda},\boldsymbol{\mu})$ by

$$h_{k,\alpha}(u,v,t;\boldsymbol{\lambda},\boldsymbol{\mu}) = \frac{\exp\left(-g_\alpha(u,v,t)\,\lambda_k + \mu_k\right)}{\sum_{w\in V_t(u)} \sum_{\ell=1}^{K} \sum_{\beta\in\mathcal{M}_{\ell,t}(u,w)} \exp\left(-g_\beta(u,w,t)\,\lambda_\ell + \mu_\ell\right)} \tag{10}$$

for $(u,v,t) \in D$, $k = 1,\ldots,K$ and $\alpha \in \mathcal{M}_{k,t}(u,v)$. From Eqs.(9) and (10), we have

$$\frac{\partial Q(\boldsymbol{\lambda}, \boldsymbol{\mu} \mid \bar{\boldsymbol{\lambda}}, \bar{\boldsymbol{\mu}})}{\partial \lambda_k} = - \sum_{(u,v,t)\in D_*} \sum_{\alpha\in\mathcal{M}_{k,t}(u,v)} \bar{q}_{k,\alpha}(u,v,t)\,g_\alpha(u,v,t) + \frac{b_k - 1}{\lambda_k} - c_k$$

$$+ \sum_{(u,v,t)\in D_*} \sum_{w\in V_{u,t}} \sum_{\alpha\in\mathcal{M}_{k,t}(u,w)} g_\alpha(u,w,t)\,h_{k,\alpha}(u,w,t;\boldsymbol{\lambda},\boldsymbol{\mu}) \tag{11}$$

$$\frac{\partial Q(\boldsymbol{\lambda}, \boldsymbol{\mu} \mid \bar{\boldsymbol{\lambda}}, \bar{\boldsymbol{\mu}})}{\partial \mu_k} = - \sum_{(u,v,t)\in D_*} \sum_{\alpha\in\mathcal{M}_{k,t}(u,v)} \bar{q}_{k,\alpha}(u,v,t) - \frac{\mu_k}{d_k^2}$$

$$+ \sum_{(u,v,t)\in D_*} \sum_{w\in V_t(u)} \sum_{\alpha\in\mathcal{M}_{k,t}(u,w)} h_{k,\alpha}(u,w,t;\boldsymbol{\lambda},\boldsymbol{\mu}) \tag{12}$$

for $k = 1,\ldots,K$. Also, from Eqs.(10), (11) and (12), we have

$$\frac{\partial^2 Q(\boldsymbol{\lambda}, \boldsymbol{\mu} \mid \bar{\boldsymbol{\lambda}}, \bar{\boldsymbol{\mu}})}{\partial \lambda_k \, \partial \lambda_\ell} = - \sum_{(u,v,t)\in D_*} \delta_{k,\ell} \sum_{w\in V_t(u)} \sum_{\alpha\in\mathcal{M}_{k,t}(u,w)} g_\alpha(u,w,t)^2\,h_{k,\alpha}(u,w,t;\boldsymbol{\lambda},\boldsymbol{\mu})$$

$$+ \sum_{(u,v,t)\in D_*} \left(\sum_{w\in V_t(u)} \sum_{\alpha\in\mathcal{M}_{k,t}(u,w)} g_\alpha(u,w,t)\,h_{k,\alpha}(u,w,t;\boldsymbol{\lambda},\boldsymbol{\mu}) \right)$$

$$\times \left(\sum_{w\in V_t(u)} \sum_{\alpha\in\mathcal{M}_{\ell,t}(u,w)} g_\alpha(u,w,t)\,h_{\ell,\alpha}(u,w,t;\boldsymbol{\lambda},\boldsymbol{\mu}) \right)$$

$$- \frac{\delta_{k,\ell}\,(b_k - 1)}{\lambda_k^2} \tag{13}$$

$$\frac{\partial^2 Q(\boldsymbol{\lambda}, \boldsymbol{\mu} \,|\, \bar{\boldsymbol{\lambda}}, \bar{\boldsymbol{\mu}})}{\partial \lambda_k \, \partial \mu_\ell} = \sum_{(u,v,t) \in D_*} \left(\sum_{w \in V_t(u)} \sum_{\alpha \in \mathcal{M}_{k,t}(u,w)} g_\alpha(u,w,t) \, h_{k,\alpha}(u,w,t; \boldsymbol{\lambda}, \boldsymbol{\mu}) \right)$$

$$\times \left(\delta_{k,\ell} - \sum_{w \in V_t(u)} \sum_{\alpha \in \mathcal{M}_{\ell,t}(u,w)} h_{\ell,\alpha}(u,w,t; \boldsymbol{\lambda}, \boldsymbol{\mu}) \right) \quad (14)$$

$$\frac{\partial^2 Q(\boldsymbol{\lambda}, \boldsymbol{\mu} \,|\, \bar{\boldsymbol{\lambda}}, \bar{\boldsymbol{\mu}})}{\partial \mu_k \, \partial \mu_\ell} = - \sum_{(u,v,t) \in D_*} \left(\sum_{w \in V_t(u)} \sum_{\alpha \in \mathcal{M}_{k,t}(u,w)} h_{k,\alpha}(u,w,t; \boldsymbol{\lambda}, \boldsymbol{\mu}) \right)$$

$$\times \left(\delta_{k,\ell} - \sum_{w \in V_t(u)} \sum_{\alpha \in \mathcal{M}_{\ell,t}(u,w)} h_{\ell,\alpha}(u,w,t; \boldsymbol{\lambda}, \boldsymbol{\mu}) \right) - \frac{\delta_{k,\ell}}{d_k^2} \quad (15)$$

for $k, \ell = 1, \ldots, K$, where $\delta_{k,\ell}$ is the Kronecker delta. We consider a quadratic form

$$\mathcal{H}_Q(\boldsymbol{x}, \boldsymbol{y}) = \sum_{k,\ell=1}^{K} \left(\frac{\partial^2 Q(\boldsymbol{\lambda}, \boldsymbol{\mu} \,|\, \bar{\boldsymbol{\lambda}}, \bar{\boldsymbol{\mu}})}{\partial \lambda_k \, \partial \lambda_\ell} x_k \, x_\ell + 2 \frac{\partial^2 Q(\boldsymbol{\lambda}, \boldsymbol{\mu} \,|\, \bar{\boldsymbol{\lambda}}, \bar{\boldsymbol{\mu}})}{\partial \lambda_k \, \partial \mu_\ell} x_k \, y_\ell + \frac{\partial^2 Q(\boldsymbol{\lambda}, \boldsymbol{\mu} \,|\, \bar{\boldsymbol{\lambda}}, \bar{\boldsymbol{\mu}})}{\partial \mu_k \, \partial \mu_\ell} y_k \, y_\ell \right)$$

$$= - \sum_{(u,v,t) \in D_*} \left(\left\langle g_\alpha(u,w,t)^2 \, x_k^2 \right\rangle - \left\langle g_\alpha(u,w,t) \, x_k \right\rangle^2 \right)$$

$$+ 2 \sum_{(u,v,t) \in D_*} \left(\left\langle g_\alpha(u,w,t) \, x_k \, y_k \right\rangle - \left\langle g_\alpha(u,w,t) \, x_k \right\rangle \left\langle y_k \right\rangle \right)$$

$$- \sum_{(u,v,t) \in D_*} \left(\left\langle y_k^2 \right\rangle - \left\langle y_k \right\rangle^2 \right) - \sum_{k=1}^{K} \left(\frac{(b_k - 1) \, x_k^2}{\lambda_k^2} + \frac{y_k^2}{d_k^2} \right) \quad (16)$$

for $\boldsymbol{x} = (x_1, \ldots, x_K)$, $\boldsymbol{y} = (y_1, \ldots, y_K) \in \mathbb{R}^K$, where $\langle z_{k,\alpha}(w) \rangle$ stands for

$$\langle z_{k,\alpha}(w) \rangle = \sum_{k=1}^{K} \sum_{w \in V_t(u)} \sum_{\alpha \in \mathcal{M}_{k,t}(u,w)} h_{k,\alpha}(u,w,t; \boldsymbol{\lambda}, \boldsymbol{\mu}) \, z_{k,\alpha}(w)$$

for $(u,v,t) \in D_*$, $k = 1, \ldots, K$, $w \in V_t(u)$ and $\alpha \in \mathcal{M}_{k,t}(u,w)$. From Eq. (10), note that

$$\sum_{k=1}^{K} \sum_{w \in V_t(u)} \sum_{\alpha \in \mathcal{M}_{k,t}(u,w)} h_{k,\alpha}(u,w,t; \boldsymbol{\lambda}, \boldsymbol{\mu}) = 1.$$

Thus, by Eq. (16), we have

$$\mathcal{H}_Q(\boldsymbol{x}, \boldsymbol{y}) = - \sum_{(u,v,t) \in D_*} \left\langle \left(g_\alpha(u,w,t) \, x_k - \langle g_\alpha(u,w,t) \, x_k \rangle - y_k + \langle y_k \rangle \right)^2 \right\rangle$$

$$- \sum_{k=1}^{K} \left(\frac{(b_k - 1) \, x_k^2}{\lambda_k^2} + \frac{y_k^2}{d_k^2} \right)$$

for $\boldsymbol{x} = (x_1, \ldots, x_K)$, $\boldsymbol{y} = (y_1, \ldots, y_K) \in \mathbb{R}^K$. This implies that the Hessian matrix of function $Q(\boldsymbol{\lambda}, \boldsymbol{\mu} \,|\, \bar{\boldsymbol{\lambda}}, \bar{\boldsymbol{\mu}})$ is negative definite. Hence, we can find the point $(\boldsymbol{\lambda}, \boldsymbol{\mu})$ at which function $Q(\boldsymbol{\lambda}, \boldsymbol{\mu} \,|\, \bar{\boldsymbol{\lambda}}, \bar{\boldsymbol{\mu}})$ attains the maximum by solving $\partial Q(\boldsymbol{\lambda}, \boldsymbol{\mu} \,|\, \bar{\boldsymbol{\lambda}}, \bar{\boldsymbol{\mu}})/\partial \lambda_k = 0$, $\partial Q(\boldsymbol{\lambda}, \boldsymbol{\mu} \,|\, \bar{\boldsymbol{\lambda}}, \bar{\boldsymbol{\mu}})/\partial \mu_k = 0$ for $k = 1, \ldots, K$. We employ Newton's method and obtain an update formula for $\boldsymbol{\lambda}$ and $\boldsymbol{\mu}$ (see Eqs. (11), (12), (13), (14) and (15)).

References

1. Barabási, A.L.: The origin of bursts and heavy tails in human dynamics. Nature **435**, 207–211 (2005)
2. Barbieri, N., Bonchi, F., Manco, G.: Who to follow and why: link prediction with explanations. In: Proceedings of KDD 2014, pp. 1266–1275 (2014)
3. Chen, W., Lakshmanan, L., Castillo, C.: Information and influence propagation in social networks. Synth. Lect. Data Manag. **5**, 1–177 (2013)
4. Crandall, D., Cosley, D., Huttenlocher, D., Kleinberg, J., Suri, S.: Feedback effects between similarity and social influence in online communities. In: Proceedings of KDD 2008, pp. 160–168 (2008)
5. Gomez-Rodriguez, M., Leskovec, J., Krause, A.: Inferring networks of diffusion and influence. In: Proceedings of KDD 2010, pp. 1019–1028 (2010)
6. Guha, R., Kumar, R., Raghavan, P., Tomkins, A.: Propagation of trust and distrust. In: Proceedings of WWW 2004, pp. 403–412 (2004)
7. Kempe, D., Kleinberg, J., Tardos, E.: Maximizing the spread of influence through a social network. In: Proceedings of KDD 2003, pp. 137–146 (2003)
8. Kimura, M., Saito, K., Nakano, R., Motoda, H.: Extracting influential nodes on a social network for information diffusion. Data Min. Knowl. Disc. **20**, 70–97 (2010)
9. Koren, Y.: Collaborative filtering with temporal dynamics. In: Proceedings of KDD 2009, pp. 447–456 (2009)
10. Leskovec, J., Backstrom, I., Kumar, K., Tomkins, A.: Microscopic evolution of social networks. In: Proceedings of KDD 2008, pp. 462–470 (2008)
11. Leskovec, J., Huttenlocher, D., Kleinberg, J.: Predicting positive and negative links in online social networks. In: Proceedings of WWW 2010, pp. 641–650 (2010)
12. Liben-Nowell, D., Kleinberg, J.: The link-prediction problem for social networks. J. Am. Soc. Inf. Sci. Technol. **58**, 1019–1031 (2007)
13. Liu, H., Lim, E., Lauw, H., Le, M., Sun, A., Srivastava, J., Kim, Y.: Predicting trusts among users of online communities: an epinion case study. In: Proceedings of EC 2008, pp. 310–319 (2008)
14. Nguyen, V., Lim, E., Jiang, J., Sun, A.: To trust or not to trust? Predicting online trusts using trust antecedent framework. In: Proceedings of ICDM 2009, pp. 896–901 (2009)
15. Oliveira, J.G., Barabási, A.L.: Dawin and Einstein correspondence patterns. Nature **437**, 1251 (2005)
16. Tang, J., Chang, S., Aggarwal, C., Liu, F.: Negative link prediction in social media. In: Proceedings of WSDM 2015, pp. 87–96 (2015)
17. Tang, J., Gao, H., Hu, X., Liu, H.: Exploiting homophily effect for trust prediction. In: Proceedings of WSDM 2013, pp. 53–62 (2013)
18. Tang, J., Gao, H., Liu, H., Sarma, A.D.: eTrust: understanding trust evolution in an online world. In: Proceedings of KDD 2012, pp. 253–261 (2012)
19. Weng, L., Ratkiewicz, J., Perra, N., Goncalves, B., Castillo, C., Bonchi, F., Schifanella, R., Menczer, F., Flammini, A.: The role of information diffusion in the evolution of social networks. In: Proceedings of KDD 2013, pp. 356–364 (2013)
20. Zhang, J., Yu, P.S., Zhou, Z.: Meta-path based multi-network collective link prediction. In: Proceedings of KDD 2014, pp. 1286–1295 (2014)

Clustering

Predicting the Primary Medical Procedure Through Clustering of Patients' Diagnoses

Mamoun Almardini[1]($\boxtimes$), Ayman Hajja[2], Zbigniew W. Raś[1,3,4], Lina Clover[5], and David Olaleye[5]

[1] College of Computing and Informatics, University of North Carolina at Charlotte,
Charlotte, NC 28223, USA
{malmardi,Ras}@uncc.edu

[2] Department of Computer Science, College of Charleston,
Charleston, SC 29424, USA
hajjaa@cofc.edu

[3] Institute of Computer Science, Warsaw University of Technology,
00-665 Warsaw, Poland

[4] Polish-Japanese Academy of Information Technology, 02-008 Warsaw, Poland

[5] SAS Institute Inc, Cary, NC 27513, USA
{Lina.Clover,David.Olaleye}@sas.com

Abstract. Healthcare spending has been increasing in the last few decades. This increase can be attributed to hospital readmissions, which is defined as a re-hospitalization of a patient after being discharged from a hospital within a short period of time. The correct selection of the primary medical procedure by physicians is the first step in the patient treatment process and is considered to be of the main causes for hospital readmissions. In this paper, we propose a recommender system that can accurately predict the primary medical procedure for a new admitted patient, given his or her set of diagnoses. The core of the recommender system relies on identifying other existing patients that are considered similar to the new patient. That said, we propose three approaches to predict the primary procedure. The results show the ability of our proposed system to identify the primary procedure. It can be later used to build a graph which shows all possible paths that a patient may undertake during the course of treatment.

Keywords: Hospital readmission · Main procedure prediction · Clustering · Personalization

1 Introduction

Recently, expenditure on healthcare has risen rapidly in the United States. According to [1], healthcare spending has been rising at twice the rate of growth of our income for the past 40 years. The projection of the growth rate in healthcare spending is 5.8% during the period 2014–2024, which means that the spending will rise to 5.4 trillion by 2024. At the same time, the gross domestic product (GDP) growth rate is only 4.7% (as of 2014) [2]. This increase in healthcare

© Springer International Publishing AG 2017
A. Appice et al. (Eds.): NFMCP 2016, LNAI 10312, pp. 117–131, 2017.
DOI: 10.1007/978-3-319-61461-8_8

spending can be attributed to several factors as listed by Price Waterhouse Coopers (PWC) Research Institute: over-testing, processing claims, ignoring doctors orders, ineffective use of technology, hospital readmissions, medical errors, unnecessary ER visits, and hospital acquired infections [3]. Figure 1 shows that 25 billion are spent annually on readmissions. Hospital readmissions and surgery outcomes prediction has gained a great interest recently in the scientific research community [4–8]. Analyzing the reasons behind readmissions and reducing them can save a great amount of money. A hospital readmission is defined as a hospitalization of the patient after being discharged from the hospital. The period in average is 30 days [7].

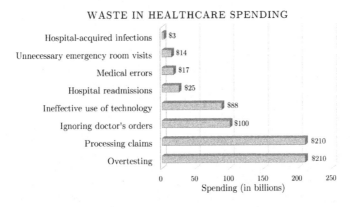

Fig. 1. Waste in healthcare spending as listed by Price Waterhouse Coopers (PWC) Research Institute [3]

One of the reasons for hospital readmissions is the wrong diagnosis of the patients. It is very important to provide the patients with the proper diagnosis in order to avoid any future readmissions and reduce the healthcare spending. In this paper, we extend our work in [9,10], where we introduced a system for physicians that recommends diagnoses transitions which would, as a result, yield to reduction in the number of anticipated hospital readmissions. The input for our system were the set of diagnoses of a new admitted patient, and the primary medical procedure assigned for that patient. However, in this research, we mine the medical dataset to predict the primary medical procedure for the patient by clustering the patients according to their set of diagnoses. We propose three new approaches to identify the patients, from the dataset, that are similar to the newly admitted patient.

2   HCUP Dataset Description

In this work, we mined the Florida State Inpatient Databases (SID) that are part of the Healthcare Cost and Utilization Project (HCUP) [11]. The SID dataset is primarily a state-level discharge data that is collected from non-federal community hospitals, which constitute the majority of hospitals in USA. The SID includes patients' demographic data, such as age, gender, and race. In addition to the demographic information, SID includes patients' medical data, such as diagnoses, procedures, status of the patient, and the length of stay. The dataset is mainly composed of three tables, namely: *American Hospital Association (AHA) Linkage, Charges*, and *Core*. The most important table in the SID is the *Core* table, which is considered as the nucleus of the SID. The *Core* table contains over 280 features (attributes); however, many of those features are repeated with different values according to the patient's status. There are two types of coding schemes used in the *Core* table for labeling and formatting, which are the International Classification of Diseases, Ninth Revision, Clinical Modification (ICD-9-CM) and the Clinical Classifications Software (CCS). The ICD-9-CM coding is detailed and uses more codes to label the procedures and diagnoses. On the other hand, CCS is more generalized and it is a collapsed form of the ICD-9-CM. For example, there are 15,072 diagnosis categories and 3,948 procedure categories in the ICD-9-CM. CCS coding however, collapses these categories into a smaller number of more generalized categories; totaling only 285 diagnoses categories, and 231 procedures categories. In the following experiments, we used the CCS coding, as it provides more meaningful and descriptive presentation of the clinical categories.

Table 1. Description of the used core table features.

Features	Concepts
VisitLink	Patient identifier
DaysToEvent	Temporal visit ordering
LOS	Length of stay
DXCCSn	n^{th} diagnosis, flexible feature
PRCCSn	n^{th} procedure, meta-action
DXPOAn	Present on admission indicator

In our experiments, we only used the features listed in Table 1 that are relevant to the examined problem. Visit linkage ($VisitLink$) feature is an encrypted identifier for patients. Each patient has a unique identifier among the hospitals within the same state. Days to event ($DaysToEvent$) feature provides information about the number of days between two consecutive visits for the same patient identified by the $VisitLink$ feature. The value of this feature is set randomly for the first visit to maintain the privacy of patients. The value of the following visit would be the initial random value assigned for the first visit plus the number of days between the admission dates of the two consecutive visits. For example, the patient can be assigned $DaysToEvent = 12$ in the first visit; which is an entirely arbitrary number and does not provide us with any information about the actual admission date. The value of $DaysToEvent$ in the second visit will be the first value (12) plus the number of days between the two visits. For example, if $DaysToEvent = 12$ for the first visit and $DaysToEvent = 40$ for the second visit, then the number of days between the two admission dates is $40 - 12 = 28$ days. It is worth mentioning here that $DaysToEvent$ represents the number of days between the admission dates, and not between the discharge date and the next admission date. $VisitLink$ and $DaysToEvent$ features are encrypted identifiers of the patients. They are used together to track patients across multiple visits within the same hospital or multiple hospitals within the same state without revealing the patient's identity. The *Length of Stay (LOS)* feature represents the number of days a patient stays at the hospital, which is the number of days from the admission date to the discharge date. $VisitLink$ and LOS can be used together to calculate the number of days between the discharge date and the next admission date. Referring back to our example above, if $VisitLink = 12$ and $LOS = 10$ for the first visit, and $VisitLink = 40$ for the second visit, then the number of days between these two visits is $40 - 12 - 10 = 18$ days. Following is the equation used to calculate the number of days between the discharge date and the next admission date:

$$DischargeToAdmissionDays = DaysToEvent_2 - DaysToEvent_1 - LOS_1 \quad (1)$$

where $DischargeToAdmissionDays$ refers to the number of days between the discharge date and the next admission date, $DaysToEvent$ refers to the number of days between any two consecutive admission dates, and LOS refers to the length of stay at the hospital. The subscript in the variable names indicates the visit number; 1 being the first visit and 2 being the second visit.

Calculating the number of days between the discharge date and the next admission date is of substantial importance, especially when the research concerns hospital readmissions. In order to consider that a patient had a readmission, the result of Eq. 1 should be less than or equal to 30 days for any two consecutive visits, as shown in Eq. 2.

$$ReadmissionIndicator = \begin{cases} Yes, & DaysBetweenDischarges \leq 30 \\ No, & DaysBetweenDischarges > 30 \end{cases} \quad (2)$$

The *Core* table reports up to 31 diagnoses $(DXCCSn)$ and up to 31 procedures $(PRCCSn)$ per discharge as it has 31 diagnosis columns and 31 procedure columns. It is worth mentioning that it is often the case that patients examination returns less than 31 diagnoses, and that the number of procedures they undergo is less than 31. Furthermore, even though a patient might have gone through several procedures during a given visit, the primary procedure that occurred at the visit discharge is assumed to be the first procedure $(PRCCS1)$. The Present on Admission $(DXPOAn)$ indicator identifies the diagnoses that were present when the patient was admitted. Since the dataset represents discharge data, then this feature is useful for identifying the diagnoses that were present at the time of admission rather than the time of discharge. In addition to the features explained above, there are several demographic data that are reported in the *Core* table as well, such as race, age range, sex, living area, etc.

3   Predicting the Primary Procedure

In the previous section, we provided a concise description of our information system (HCUP), in which each instance (or visit) consists of one primary procedure and a set of diagnoses; when a new patient is admitted to the hospital, the physicians examine his or her set of diagnoses and assign a primary procedure accordingly. In [9], Almardini et al. introduced a system for physicians that recommends diagnoses transitions which would, as a result, yield to reduction in the number of anticipated hospital readmissions. The input for their system were the set of diagnoses of a new admitted patient, and the primary procedure assigned for that patient. The recommender system presented in [9] however, was not built to provide any recommendations on what the primary procedure should be. In this paper, we examine few approaches that address the challenge of predicting the primary procedure for a patient, given his or her set of diagnoses.

The goal of our system, which is to accurately predict the primary procedure for a newly admitted patient, is almost wholly determined by its ability to identify other existing patients that are considered similar to our admitted patient. The basis for determining similarities between different patients however, which we will explore next, is an intricate endeavor, given that the input of our patients is a set of diagnoses that differ greatly in the level of significance. Figure 2 shows the architecture of the proposed system.

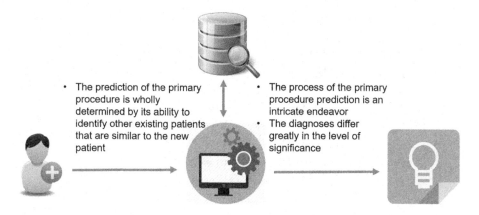

- The prediction of the primary procedure is wholly determined by its ability to identify other existing patients that are similar to the new patient
- The process of the primary procedure prediction is an intricate endeavor
- The diagnoses differ greatly in the level of significance

Fig. 2. The architecture of the system

3.1   Minimum Similarity Match

The first approach that we propose to predict the primary procedure, is to have our similarity function be defined in a way that marks a newly admitted patient (p_n) similar to an existing patient (p_e) if and only if the existing patient exhibits every single diagnoses present in the admitted patient:

$$\text{similarity}(p_n, p_e) = \begin{cases} 1, & \text{if } diag(p_n) \subset diag(p_e) \\ 0, & \text{otherwise} \end{cases} \tag{3}$$

where 1 indicates that the new patient (p_n) is similar to the existing patient (p_e), and 0 otherwise. Consequently, we can define A_n as the set of all existing patients that the new patient (p_n) is similar to:

$$A_n = \bigcup_{i=1}^{m} \{p_i : \text{similarity}(p_n, p_i) = 1\} \tag{4}$$

where m is the number of existing patients in our dataset who are similar to p_n, and p_i is the ith existing patient in that set. The final output for our recommender system is a probability distribution for the primary procedures obtained by the set of all similar existing patients (A_n). To demonstrate with an example, say we have a newly admitted patient (p_n) with the following set of diagnoses: $diag(p_n) = \{d1, d3, d5\}$, let us also assume that our dataset consists of the set of seven patients shown in Table 2.

Looking at our dataset of patients in Table 2, we can conclude the following:

- similarity$(p_n, p_i) = 0$, for i $= 1$ and 4
- similarity$(p_n, p_i) = 1$, for i $= 2$, 3, 5, 6, and 7

Table 2. Dataset S, containing all existing patients

	Diagnoses	Primary procedure
p_1	{d1, d2, d5, d8}	Procedure 6
p_2	{d1, d2, d3, d5}	Procedure 3
p_3	{d1, d3, d4, d5, d9}	Procedure 6
p_4	{d1, d2, d3, d6}	Procedure 2
p_5	{d1, d3, d4, d5, d8}	Procedure 3
p_6	{d1, d2, d3, d4, d5}	Procedure 2
p_7	{d1, d3, d5, d6, d7}	Procedure 3

According to our previous definitions, A_n will contain the set of elements: p_2, p_3, p_5, p_6, and p_7 and the primary procedures for these patients are 3, 6, 3, 2, and 3 respectively. Therefore, the output to our recommender system will be 60% Procedure 3, 20% Procedure 6, and 20% Procedure 2, which is the probability distribution of the primary procedures of A_n.

Table 3 shows a list of the accuracies for our system when tested on 815 randomly selected instances, each being compared to roughly 4 millions existing patients using our definition of similarity presented earlier. As can be seen in Table 2, the procedure with the highest probability in the existing matches distribution was predicted correctly 18.5% of the time, the correct primary procedure was predicted correctly as one of the two procedures 23.6% of the time, and the correct primary procedure was predicted correctly as one of the three procedures 26.5% of the time. The frequency is the number of instances, out of the 815, for which the primary procedure was predicted correctly.

Table 3. Prediction accuracy of the minimum similarity match using the N most probable primary procedures

N	Frequency	Accuracy
1	152	18.5%
2	192	23.6%
3	216	26.5%
4	233	28.6%
5	243	29.8%
6	250	30.7%
7	260	31.9%
8	265	32.5%
9	268	32.9%
10	270	33.1%

Although the approach presented in this section is showing reasonably good results, the fact that our definition of similarities requires an existing patient (p_e) to exhibit all diagnoses of the new patient (p_n) makes this system rather limited. Therefore, we need to apply a more flexible system for identifying similarities that can increase the number of patients that are similar to the newly admitted patient.

3.2   Jaccard Similarity Match

According to our dataset, a patient has on average 7.55 diagnoses presented on admission. In addition, there are 30.99%, 10.67%, and 2.68% of patients having 10 diagnoses or more, 15 diagnoses or more, and 20 diagnoses or more, respectively. Therefore, there is a high probability that a newly admitted patient will exhibit a large number of diagnoses, which would make it hard to identify similar patients in our dataset using the algorithm presented in Sect. 3.1. One way to tackle this limitation is to modify the definition of the similarity between patients in a way that when a new patient gets admitted to the hospital, rather than finding patients who have all the diagnoses that the new patient has, we measure the percentage of diagnoses that are common between the new patient and the existing patients in the dataset. This can be achieved by applying Jaccard similarity index, which measures the similarity between two sets by dividing the cardinality of the intersection by the cardinality of the union as described in the following equation:

$$J(A, B) = \frac{|A \cap B|}{|A \cup B|} \qquad (5)$$

where A and B represent the two sets of diagnoses, and where $|.|$ represents the cardinality of the intersection and union of the two sets.

The elements of the sets are the diagnoses of the patients. One of the sets contains the new patient's diagnoses and the other set contains the existing patient's diagnoses. In our dataset, we calculate Jaccard similarity of the new patients with every patient in the dataset. The patient with the highest Jaccard index is considered the most similar to the new patient. Now that we have identified the patient that is most similar to our newly admitted patient, we can consequently use the existing patient's main procedure to predict the main procedure for our new patient. However, doing so is rather unfavorable, since building an entire prediction system based upon one patient, may lead to a bias in our results. Therefore, it would be more robust to instead consider all patients that satisfy a similarity value within a given margin (e.g. x top most similar patients). In our implementation, we have tested using different margins and found that 5% would yield the best results. For example, suppose we have a dataset of 40 patients and the similarities of these patients with the newly admitted patient fall in this range [40–90%]. Given a margin of 5%, we would select the top 2 similarities out of the 40 similarities; which means that we will select the primary procedures of the patients with the two highest similarities.

The best way to explain how this algorithm works is by providing an example. Suppose a new patient pat_0 comes to the hospital with a certain number of diagnoses $\{d_1, d_3, d_5\}$. Also, suppose that our dataset contains 10,000 patients as listed in Table 4. The first step would be to find the intersection and union of p_0 with all the patients in the dataset, then calculate the Jaccard index. For example, the similarity index of pat_0 and pat_1 is calculated as follows:

$$J(pat_0, pat_1) = \frac{|pat_0 \cap pat_1|}{|pat_0 \cup pat_1|} = \frac{|\{d_1, d_5\}|}{|\{d_1, d_2, d_3, d_5, d_8\}|} = 2/5 = 0.4 \qquad (6)$$

As seen in Table 4, the range of the similarities is [40–75]. Given this range, we would select the highest 5% similarities and store the primary procedures associated with them. After that, we find the average similarity of each procedure. So, assuming that the only patients with most frequent primary procedure p_6 that are within our specified margins are pat_2 and pat_3, then the average similarity index for p_6 would be the average of 75% and 60%, which is equal to 67.5%; and the average similarity index of p_3 would be 75%, assuming that pat_4 is the only patient that exhibits p_3 whom also lies within our specified similarity margin. Finally, our prediction will be the procedure with the highest average (weight). We have run Jaccard similarity algorithm on a dataset of 10,000 patients and measured the accuracy using 10-fold cross validation. The resulted accuracy was

20.25%, which is slightly better than the accuracy of the Minimum Similarity Match (MSM) method used in Sect. 3.1.

Table 4. Jaccard similarity calculations example

Patient	Diagnoses	Most frequent primary procedure	Similarity index
pat_1	$\{d_1, d_2, d_5, d_8\}$	p_6	40%
pat_2	$\{d_1, d_2, d_3, d_5\}$	p_6	75%
pat_3	$\{d_1, d_3, d_4, d_5, d_9\}$	p_6	60%
pat_4	$\{d_1, d_3, d_4, d_5\}$	p_3	75%
...	...	...	...
pat_{9999}	$\{d_1, d_2, d_3, d_6\}$	p_2	40%
pat_{1000}	$\{d_1, d_3, d_4\}$	p_3	50%

The approaches presented in Sects. 3.1 and 3.2 are showing reasonably good results in predicting the primary procedure. In these approaches however, we based our definition of the patients' similarity on the number of diagnoses that the patients exhibit. However, we should shift our focus to the level of importance of each diagnoses with respect to their abilities to predict the primary procedure. There is typically only a small number of subsets that are capable of determining the primary procedure. In the next subsection, we present a new and novel approach on how to identify such sets.

3.3 Selective Similarity Match

In this subsection, we introduce an enhanced system for predicting the primary procedure for new patients. Our approach presented here is based on the fact that there is only a selected number of combinations for diagnoses subsets that are capable of predicting primary procedures. This means that for a new patient exhibiting x number of diagnoses, it would be more likely the case that matching our dataset for patients that exhibit only a subset of the x diagnoses will yield better result; by doing so, our system will not only avoid overfitting, but it will also result in many more matches in our existing dataset and the same in a higher level of prediction accuracy. The level of predictability for a subset of diagnoses s, can be determined based on the distribution of the primary procedures for existing patients that exhibit s. By calculating the entropy of the main procedures for each possible subset of the diagnoses, we are able to identify subsets that can most accurately predict the primary procedure (subsets that have the least entropy values).

Our system starts by generating all possible combinations of k-diagnosis sets, starting with k = 1 and ending with k = 3, then calculating the entropy of the primary procedures for each combination. For each combination of diagnoses s, we identify all existing patients that belong to s, then we calculate the entropy $H(s)$ according to the distribution of the primary procedures for s:

$$H(s) = -\sum_{i=1}^{m} p_i \log(p_i) \qquad (7)$$

where p_i is the probability of the ith primary procedure, and m is the number of primary procedures in s.

The reason for why we stop at the number 3 is because the number of distinct subsets that can be generated from the set of all 285 diagnoses grows exponentially large as k increases. For example, the number of unique 3-diagnoses subsets that can be chosen from 285 diagnoses is roughly 4 millions; the number of unique 4-diagnoses subsets however, exceeds 250 millions.

For a new admitted patient with x number of diagnoses, we generate all subsets of k-diagnoses for $k = 1, 2,$ and 3; then, using our previously calculated entropies for all possible diagnoses, we identify the subset of the patient diagnoses with the lowest entropy (highest level of predictability), and use its most frequent procedure as the anticipated primary procedures. We have run the Selective Similarity Match (SSM) algorithm on a dataset of 10,000 patients and measured the accuracy using 10-fold cross validation. The resulted accuracy was 25.25%, which is better than both the accuracies of the Minimum Similarity Match (MSM) and Jaccard Similarity Match approaches used in Sects. 3.1 and 3.2 respectively.

Next, we provide a real example from our dataset to demonstrate the algorithm.

Let us first assume that the first step of the algorithm, which is to generate all possible combinations of k-diagnosis sets, starting with k = 1 and ending with k = 3 has been performed. Now, say that a new patient (p_n) has been admitted to the hospital with the following set of diagnoses {181, 183, 101, 164}:

– 181: Other complications of pregnancy.
– 183: Hypertension complicating pregnancy; childbirth and the puerperium.
– 184: Early or threatened labor.
– 189: Previous C-section.

The next step would be to generate all 1-diagnosis, 2-diagnoses, and 3-diagnoses subsets of (p_n), which is shown in the first column of Table 5.

According to Table 5, the list of diagnoses that has the least entropy is {181, 183, 189}, in which the most probable primary procedure is 134 (Cesarean section), which is indeed the correct primary procedure for our patient (p_n). Following is a description of the procedure codes found in Table 5:

– 134: Cesarean section.
– 137: Other procedures to assist delivery.

Table 6 shows few procedures with their prediction accuracy for a testing sample of 1,000 instances, using a training set of size 10,000 instances. For example,

Table 5. An example of one of the tested patients

List of diagnoses in cluster	Entropy	Primary procedure
181	2.414	137
183	2.258	137
184	1.564	137
189	1.293	134
181, 183	2.419	137
181, 184	1.783	137
181, 189	1.224	134
183, 184	1.241	134
183, 189	0.622	134
184, 189	0.884	134
181, 183, 184	-	-
181, 183, 189	0.337	134
181, 184, 189	1.095	134
183, 184, 189	-	-

Table 6. Sample of main procedures with their frequencies and accuracies for a testing sample of 1,000 instances, using a training set of size 10,000 instances

Procedure	Frequency	Accuracy
137 (Other Procedures to Assist Delivery)	23	65%
84 (Cholecystectomy and Common Duct Exploration)	21	62%
158 (Spinal Fusion)	10	60%
152 (Arthroplasty Knee)	14	57%
134 (Cesarean Section)	38	45%
61 (Other OR Procedures on Vessels Other than Head and Neck)	18	44%
45 (Percutaneous Transluminal Coronary Angioplasty PTCA)	45	42%
78 (Colorectal Resection)	32	39%
124 (Hysterectomy Abdominal and Vaginal)	19	32%
70 (Upper Gastrointestinal Endoscopy Biopsy)	148	31%
47(Diagnostic Cardiac Catheterization Coronary Arteriography)	27	30%

the third row in our table states that we encountered 10 instances (out of our 1,000 testing sample) with main procedure 'Spinal Fusion', and that we were able to predict this procedure with accuracy 60%, meaning that we were able to correctly predict that the main procedure is *Spinal Fusion*, for 6 instances out of 10.

4   Introducing Procedure Paths and Procedure Graph

In previous sections, we have shown how to predict the primary procedure using different approaches. Given the prediction of the primary procedure for a newly admitted patient, we can also predict the *procedure path*; which is defined as the sequence of procedures that a given patient undertakes to reach a desired treatment. In other words, a procedure path is a detailed description for the course of treatments provided to an admitted patient. The length of any given procedure path is an indicator of the number of readmissions that occurred or will occur throughout the course of treatment. For example, one procedure path for a patient could be the following: $path_x = (p_1, p_3, p_3, p_6)$, where p_i $(i = 1, 3, 6)$ indicates a particular procedure; according to procedure path $path_x$, the number of readmissions was 3.

The *procedure graph* for some procedure p is defined as the tree of all possible procedure paths extracted from our dataset for patients who underwent procedure p as their first procedure. Figure 3 shows a depiction of the *procedure graph*; $P_{(0,1)}$ is the initial procedure that patients start with; the next procedure could be any procedure from $P_{(1,1)}$ to $P_{(1,n)}$, which is determined by the resulting set of diagnoses after performing the initial procedure $P_{(0,1)}$. The first argument x in the notation $P_{(x,y)}$ refers to the number (or rather level) of readmissions, and the second argument y refers to the procedure identifier at that level. For example, $P_{(1,2)}$ refers to the procedure with identifier 2 that occurred at the first level of readmissions (e.g. first readmission following the initial procedure). The portions of the graph that are contained in dashed boxes depict the personalization part that we introduce in the next section. The idea of personalization is to cluster patients that are scheduled to undergo procedure $P_{(0,1)}$ according to their diagnoses; as a result of this clustering, we will be able to anticipate with higher accuracy the following procedure (readmission) that the patient will undergo by identifying which cluster the new patient belongs to. In the following sections, we will provide some information about the number of different possible paths and the length of each path.

4.1   Unique Procedure Paths

The number of all procedure paths is extremely high. This high number of unique procedure paths indicates that it is not true that there exists a single universal course of treatment that patients typically follow to reach the desired state. For example, the number of patients that underwent procedure 222 (Blood Transfusion), as their first procedure, is 72,521 and the number of unique procedure paths that those patients underwent is 1,230 paths. Blood transfusion is considered a minor procedure and this could explain the high number of unique procedure paths. Now, let us consider a major procedure, such as 158 (Spinal fusion), the number of patients who underwent this procedure, as their first procedure, is 72,928 and the number of unique procedure paths that those patients underwent is 443 paths. Although the number of unique paths, in case of the major procedure, is reduced by 1/3. However, the number is still high and this emphasizes the fact that patients do not follow the same path in their course of treatment.

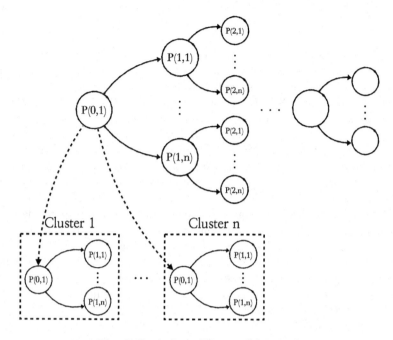

Fig. 3. Depiction of a procedure graph

4.2 Lengths of Procedures Paths

The length of the procedure path is an indicator of the number of readmissions during the course of treatment. A procedure may have different paths' lengths depending on the number of possible unique paths that the patient may follow. Knowing the length of the path is a valuable information for the physicians in the treatment process. The fact that the physicians can now anticipate the next procedure in the course of treatment makes them reconsider their decisions and select a better path that has a shorter length. The relation between the length of the path (number of readmissions) and the number of nodes (procedures) in that path is defined as follows:

Path length (Number of readmissions) = Number of nodes (procedures) − 1

Let us now consider a real example from the dataset showing the different paths' lengths for a certain procedure. Suppose that a new patient is admitted to the hospital and our system predicted that procedure 58 (Hemodialysis) to be the initial procedure based on the patient's diagnoses. Knowing that the patient will undertake procedure 58 as the initial procedure allows us to anticipate what could be the maximum and the average number of possible readmissions; which are in this case 36 and 4.2 respectively.

5   Conclusion

Predicting the primary medical procedure for a new patient is of great help to physicians; since it gives them confidence of their medical decisions and helps to achieve the desired outcomes. In this research, we proposed a recommender system that can accurately predict the primary medical procedure for a newly admitted patient through finding the similarity with the old patients according to their set of diagnoses. We proposed three approaches, namely: Minimum Similarity Match (MSM), Jaccard Similarity Match (JSM) and Silictive Similarity Match (SSM) to address the challenge of predicting the primary procedure for a patient, given his or her set of diagnoses. The three approaches showed a high level of predictability. However, the third approach is more accurate due to its ability of identifying the significant diagnoses that are responsible for the patient's admission. The predicted primary procedure is then used to anticipate all possible paths that a patient may undertake during the course of treatment; which plays a significant role in the medical decision making process. As a future work, we are planning to mine the procedures that are medically associated with the primary procedure and then recommend a set of procedures according to the patient's demographic and medical details.

Acknowledgment. This work was supported by SAS Institute under UNC-Charlotte Internal Grant No. 15-0645.

References

1. Goodman, J.C.: Priceless: Curing the Healthcare Crisis. Independent Institute, Oakland (2012)
2. Keehan, S.P., et al.: National health expenditure projections, 2014–24: spending growth faster than recent trends. Health Aff. **34**(8), 1407–1417 (2015)
3. Price Waterhouse Coopers: The Price of Excess. Identifying Waste in Healthcare Spending. Price Waterhouse Coopers Research Institute, London (2006)
4. Touati, H., Raś, Z.W., Studnicki, J., Wieczorkowska, A.A.: Mining surgical meta-actions effects with variable diagnoses' number. In: Andreasen, T., Christiansen, H., Cubero, J.-C., Raś, Z.W. (eds.) ISMIS 2014. LNCS, vol. 8502, pp. 254–263. Springer, Cham (2014). doi:10.1007/978-3-319-08326-1_26
5. Lally, A., Bachi, S., Barborak, M.A., Buchanan, D.W., Chu-Carroll, J., Ferrucci, D.A., Glass, M.R., Kalyanpur, A., Mueller, E.T., Murdock, J.W., Patwardhan, S.: WatsonPaths: scenario-based question answering and inference over unstructured information. Technical report, Research report RC25489, IBM Research (2014)
6. Tremblay, M.C., Berndt, D.J., Studnicki, J.: Feature selection for predicting surgical outcomes. In: Proceedings of 39th Annual Hawaii International Conference on System Sciences, HICSS 2006, vol. 5, p. 93a. IEEE (2006)
7. Silow-Carroll, S., Edwards, J.N., Lashbrook, A.: Reducing hospital readmissions: lessons from top-performing hospitals. CareManagement **17**(5), 14 (2011)
8. Miotto, R., Li, L., Kidd, B.A., Dudley, J.T.: Deep patient: an unsupervised representation to predict the future of patients from the electronic health records. Scientific reports 6 (2016)

9. Almardini, M., Hajja, A., Raś, Z.W., Clover, L., Olaleye, D., Park, Y., Paulson, J., Xiao, Y.: Reduction of readmissions to hospitals based on actionable knowledge discovery and personalization. In: Kozielski, S., Mrozek, D., Kasprowski, P., Małysiak-Mrozek, B., Kostrzewa, D. (eds.) BDAS 2015-2016. CCIS, vol. 613, pp. 39–55. Springer, Cham (2016). doi:10.1007/978-3-319-34099-9_3
10. Almardini, M., Hajja, A., Clover, L., Olaleye, D., Park, Y., Paulson, J., Xiao, Y.: Reduction of hospital readmissions through clustering based actionable knowledge mining. In: Proceedings of IEEE/WIC/ACM International Conference on Web Intelligence (WI 2016), pp. 444–448. IEEE Computer Society (2016)
11. Healthcare Cost and Utilization Project (HCUP). Clinical classifications software (ccs). http://www.hcup-us.ahrq.gov

Feature Clustering for Extreme Events Analysis, with Application to Extreme Stream-Flow Data

Maël Chiapino[(⊠)] and Anne Sabourin

LTCI, Télécom ParisTech, Université Paris-Saclay, Paris, France
mael.chiapino@telecom-paristech.fr

Abstract. The dependence structure of extreme events of multivariate nature plays a special role for risk management applications, in particular in hydrology (flood risk). In a high dimensional context ($d > 50$), a natural first step is dimension reduction. Analyzing the tails of a dataset requires specific approaches: earlier works have proposed a definition of sparsity adapted for extremes, together with an algorithm detecting such a pattern under strong sparsity assumptions. Given a dataset that exhibits no clear sparsity pattern we propose a clustering algorithm allowing to group together the features that are 'dependent at extreme level', i.e.,that are likely to take extreme values simultaneously. To bypass the computational issues that arise when it comes to dealing with possibly $O(2^d)$ subsets of features, our algorithm exploits the graphical structure stemming from the definition of the clusters, similarly to the Apriori algorithm, which reduces drastically the number of subsets to be screened. Results on simulated and real data show that our method allows a fast recovery of a meaningful summary of the dependence structure of extremes.

Keywords: Extreme values · Dimension reduction · Pattern mining · Subspace clustering · Subgroup discovery

1 Introduction

Extreme value analysis is of primarily interest in many contexts. One example is the machine learning problem of anomaly detection, where one needs to control the false positive rate in the most remote regions of the sample space [7,16, 17,21]. Another example is the field of environmental sciences, where extreme events (floods, droughts, heavy rainfall, . . .) are of particular concern to risk management, considering the disastrous impact these events may have. Using Extreme Value Theory (EVT) as a general setting to understand or predict extreme events has a long history [20]. In spatial problems, exhibiting areas (groups of weather stations) which may be concomitantly impacted by severe events is of direct interest for risk management policies. Identifying these groups may also serve as a preliminary dimensionality reduction step before more precise modeling. Before proceeding further, we emphasize that standard dimension

© Springer International Publishing AG 2017
A. Appice et al. (Eds.): NFMCP 2016, LNAI 10312, pp. 132–147, 2017.
DOI: 10.1007/978-3-319-61461-8_9

reduction techniques such as PCA do not apply to extremes as these methods essentially focus on the data around the mean by analyzing their covariance structure, which does not characterize the behavior of extremes (i.e., data far away in the tails of the distribution). In the present paper, the quantity of interest is river water-flow recorded at several locations of the French river system. The features of the experiment are thus the stream-flow records at different gauging stations, and the goal is to recover maximal groups of stations where extreme discharge may occur simultaneously. Our dataset consists of daily stream-flow recorded at 92 gauging stations scattered over the French river system, from 1969, January 1st to 2008, December 31st. It is the same dataset as in [14], up to 220 gauging stations presenting missing or censored records, which have been removed from our analysis, which results in $n = 14610$ vectors $X_1, .., X_n$ in $\mathbb{R}^d$, with $d = 92$ the number of stations. The reader is referred to [14] for more details.

Related Work. Dimensionality reduction for extreme value analysis has emerged very recently in the literature. As far as we know, the seminal contribution is [6] and is restricted to moderate dimensional settings ($d \leq 20$, see Sect. 3.1 for more details). The methodology proposed by [6] allows to recover groups of components (features) which may take large values simultaneously, *while the other features stay small.* For the purpose of anomaly detection, [16,17] proposed an alternative algorithm to do so with a reduced computational complexity of order $O(nd \log n)$. To the best of our knowledge, these are currently the only available examples in the literature to handle the recovery of groups of features which are representative of the *extremal dependence structure.* (See Sect. 2.2 for a precise definition of the latter). In [16,17], the extremal dependence structure is called *sparse* if the number of such groups is small compared with $2^d - 1$, the total number of groups. The output of [16,17]'s DAMEX algorithm is a (hopefully sparse) vector $\hat{\mathcal{M}} = (\hat{\mu}_\alpha, \alpha \subset \{1, \ldots, d\})$ of size $2^d - 1$, where $\hat{\mu}_\alpha$ is a summary of the dependence strength at extreme levels between features $j \in \alpha$. The fact that $\hat{\mu}_\alpha$ is positive means that the probability that all features in α be large while all others stay small, is not negligible. Various datasets have been analyzed in [16,17] (wave data from the north sea, standard anomaly detection datasets, simulated data) for which the DAMEX algorithm does exhibit a sparsity pattern, thus pointing to a relatively small number of groups of features α (each being of relatively small size $|\alpha|$ compared to the original dimension of the problem) which could be jointly extreme. However, DAMEX becomes unusable in situations where the subsets of features impacted by extreme events vary from one event to another: DAMEX then finds a very large number of subsets to be dependent, but not significantly so, (i.e., $0 < \hat{\mu}_\alpha \ll 1$), so that no sparsity pattern emerges. This is precisely the case with the river flow dataset analyzed in the present paper (see Sect. 5).

Contributions. One remarkable aspect of the preliminary analysis of the river flow dataset using DAMEX is the tendency of those many subsets α's such that $\hat{\mu}_\alpha > 0$, to form *clusters,* whose members differ from each other by a single or two features only. In practice, this means that several distinct events have impacted 'almost' the same group (cluster) of stations. The aim of this paper is to propose a methodology enabling to gather together such 'close-by' feature subsets into feature clusters. This is done by relaxing the constraint that 'features not in α take small values' when constructing the representation of the dependence structure. The output of the CLEF algorithm (CLustering Extreme Features) proposed in the present work (Sect. 4) is an alternative representation which remains usable in this 'weakly sparse' context. This representation can still be explained and understood in the multivariate EVT framework (Sect. 3), as in [6,16,17]. We emphasize that the scope of CLEF algorithm concerns situations similar to the hydrological problem considered here, where the DAMEX algorithm does not yield a readable output. In the opposite case (*e.g.* with the wave dataset or the anomaly detection datasets analyzed in [16,17]), DAMEX remains a better option than CLEF in view of its computational simplicity.

Relationships with Apriori. The dimension reduction problem considered here (determining for which subgroups of features concomitant large values are frequent) is closely related to the problem of frequent itemsets mining, specifically to the well known Apriori algorithm introduced by [2], see also [19]. Indeed, the present problem can be recast as follows: encoding as a '1' any value above a specified threshold and as a '0' any value below this threshold, one obtains a binary dataset. The goal is now to recover the groups items (features) for which concomitant '1' values are frequent, which is precisely the frequent itemsets mining problem. The combinatorial issue that arises with possibly $2^d - 1$ subsets is circumvented in Apriori by considering subsets of increasing sizes, letting a subset 'grow' until its frequency in the database is not significant anymore. This incremental principle is also related to a subset clustering method proposed in [1]. CLEF proceeds in a similar way to Apriori, the main difference being that CLEF comes with a natural interpretation in terms of multivariate EVT. Also, in practice, the stopping criterion used to decide whether incrementing a feature subset is different in CLEF and in Apriori, allowing CLEF to detect larger groups, as discussed in Sects. 3.2 and 4.1.

The paper is organized as follows. Section 2 sets up the extremal feature clustering problem and establishes connections with multivariate EVT. The dimension reduction method that we promote is explained in Sect. 3: Sect. 3.1 recalls existing work and points out some limitations, Sect. 3.2 makes explicit the links between the considered problem and the Apriori algorithm. The CLEF algorithm is described in Sect. 4. Section 5 gathers results: the output of CLEF is compared with that of DAMEX and Apriori. Section 6 concludes. The Python code for CLEF, the scripts and the dataset used for our hydrological case study are available at https://bitbucket.org/mchiapino/clef_algo.

2 Problem Statement and Multivariate EVT Viewpoint

2.1 Formal Statement of the Problem

Consider a random vector $X = (X^1, \ldots, X^d)$ in $\mathbb{R}^d$ (here, X^j is the water discharge recorded at location j). The first step when it comes to learning dependence properties of X is to standardize the features, in the same spirit as in the copula framework, which allows one to focus only on the *dependence* structure of X. One popular standardization choice in multivariate EVT is the probability integral transform: Denote by F the joint cumulative distribution function (*c.d.f.*) of X and by F^j the marginal *c.d.f.* of X^j. For simplicity, let us assume that each F^j is continuous (no point masses), so that with probability one, $0 < F^j(X^j) < 1$. The standardized variable used for dependence analysis are $V^j = (1 - F^j(X^j))^{-1}$, $j = 1, \ldots, d$ and $V = (V^1, \ldots, V^d)$. Doing so, the V^j's are identically distributed according to standard Pareto distribution, $\mathbb{P}(V^j > t) = 1/t, t \geq 1$. Our goal here is to recover all the maximal subsets of features (stations) $\alpha \subset \{1, \ldots, d\}$ which 'may be large together' with non negligible probability. In more formal terms, define the *extremal joint excess coefficient*,

$$\rho_\alpha := \lim_{t \to \infty} t\mathbb{P}\left(\forall j \in \alpha, V^j > t\right) = \lim_{t \to \infty} \mathbb{P}\left(\forall j \in \alpha, V^j > t \mid V^{\alpha_1} > t\right) \in [0, 1]. \quad (1)$$

The variable t plays the role of a high threshold above which the standardized feature V^j is considered as extreme. In practice, estimation will be done by fixing a large t and assuming that the limit in (1) is approximately reached. An advantage of the standardization procedure is that a single threshold t is needed to define an extreme event, not d thresholds, since all the features share the same scale. The limit in (1) exists under the regularity property (3) in the next paragraph. Notice already that the second equality also comes from our standardization choice ensuring that for any $j \leq d$, $t^{-1} = \mathbb{P}(V^j > t) = \mathbb{P}(V^{\alpha_1} > t)$, which justifies the scaling factor t in the definition. The coefficient $\rho_\alpha \in [0, 1]$ may be seen as a 'correlation' coefficient for the features $X^j, j \in \alpha$ at extreme levels. We say that the features $\{V^j, j \in \alpha\}$ 'may be large together' if $\rho_\alpha > 0$. One relevant summary of the dependence structure of extremes is thus the set of subgroups

$$\mathbb{M} = \{\alpha \subset \{1, \ldots, d\} : \quad \rho_\alpha > 0\}. \quad (2)$$

More precisely, we would like to recover those subgroups $\alpha \in \mathbb{M}$ which are maximal for inclusion in $\mathbb{M}$, i.e., $\forall \beta$ such that $\alpha \subsetneq \beta$, $\beta \notin \mathbb{M}$. A maximal set of features $\alpha \in \mathbb{M}$ may be viewed as a *cluster*, in the sense that every subset $\beta \subset \alpha$ is dependent at extreme level (i.e., $\rho_\beta > 0$), and that α 'gathers' all of them together. In this paper, a 'cluster' of features is understood as a maximal element $\alpha \in \mathbb{M}$.

2.2   Connections with Multivariate EVT

The working hypothesis in EVT is that, up to marginal standardization, the distribution of X is 'approximately homogeneous' on extreme regions. As pointed out above, if the margins F^j are continuous, then the V^j's have the homogeneity property: $t\mathbb{P}\left(\frac{V^j}{t} \geq x\right) = 1/x$, for $1 \leq j \leq d, t > 1, x > 0$. The key assumption is that the latter property holds *jointly* at extreme levels, i.e., that V is jointly *regularly varying* (see e.g. [23]), which writes

$$t\mathbb{P}\left(\frac{V}{t} \in A\right) \xrightarrow[t \to \infty]{} \mu(A), \quad (3)$$

where μ is the so-called *exponent measure* and where A is any set in $\mathbb{R}^d$ which is bounded away from 0 and such that $\mu(\partial A) = 0$. The exponent measure is finite on any such set A and satisfies, for $t > 0, A \subset \mathbb{R}^d_+$, $t\mu(tA) = \mu(A)$, where $tA = \{tx : x \in A\}$. Notice that many commonly used textbook multivariate distributions (*e.g.* multivariate Gaussian or Student distributions) satisfy (3), after standardization to V variables. The measure μ characterizes the distribution of V at extreme levels, since for t large enough (so that the region tA is an 'extreme region' of interest), one may use the approximation $\mathbb{P}(V \in tA) \simeq t^{-1}\mu(A)$. The connection between μ and the ρ_α's is as follows: consider the 'rectangle'

$$\Gamma_\alpha := \{x \in \mathbb{R}^d_+ : \quad \forall j \in \alpha, \ x^j > 1\} \tag{4}$$

From the definitions (1) and (3), it follows that $\rho_\alpha = \mu(\Gamma_\alpha)$. Thus the family of subset $\mathbb{M}$ in (2) writes $\mathbb{M} = \{\alpha : \mu(\Gamma_\alpha) > 0\}$.

Non Parametric Estimation. In a word, non parametric estimation of extremal characteristics based on *i.i.d.* data $X_1, \ldots, X_n$ (distributed as X) is performed by replacing probability distributions with their empirical counterparts, and by proceeding as if the limit in (3) were reached above some large threshold t. Since the F^j's are unknown, set $\hat{V}_i^j = 1/(1 - \hat{F}^j(X_i^j))$, $= 1, \ldots, n$, $j = 1, \ldots, d$, where $\hat{F}^j(x) = n^{-1}\sum_{i=1}^n \mathbb{1}\{X_i^j < x\}$. Thus $\hat{V}_i^j \in \{1, n/(n-1), n/(n-2), \ldots, n/2, n\}$ and for each fixed j, and $t \leq n$, the number of examples i such that $\hat{V}_i^j > t$ is equal to $\lceil n/t \rceil$. This suggests that t should be chosen as a function of the sample size and indeed, theoretical guarantees on the estimators are obtained for $t = o(n)$ and $t \to \infty$, *e.g.* $t \sim \sqrt{n}$, see [3], Chap. 3 for more details. After this data preprocessing step, the exponent measure μ of any region $A \subset \mathbb{R}^d_+ \setminus \{0\}$ is approximated by

$$\mu_n(A) = t\hat{P}_n(tA), \qquad \text{where} \quad \hat{P}_n(A) = n^{-1}\sum_{i=1}^n \delta_{\hat{V}_i}(A), \tag{5}$$

where δ denotes the Dirac mass. Statistical properties of μ_n (or of other functional summaries of it) have been investigated by many authors, see *e.g.* [11,12,22] for the asymptotic behavior, [15] for finite sample error bounds.

3   Dimension Reduction for Multivariate Extremes

3.1   Existing Work

Numerous modeling strategies for low dimensional multivariate extremes (say $d \leq 10$) have been proposed, see *e.g.* [9,10,25] for parametric modeling, [4,13,18,24] for semi- or non-parametric ones. For higher dimensional problems, to this date, the only available dimensionality reduction methods are (to our best knowledge) the recent works [6,16,17]. These three references share the common idea of recovering the sub-cones of $\mathbb{R}^d_+$ on which the exponent measure μ concentrates. The seminal paper on this subject appears to be [6]. It relies on principle nested spheres and spherical k-means and is designed for moderate dimensional problems only ($d \leq 20$ in their simulation experiments and $d = 4$ in their case study) with a relatively simple dependence structure (at most 4 groups of features with extremal dependence, only two for $d = 20$ in their simulation experiments). The computational burden significantly increases for larger dimensions or more elaborate dependence structures, as discussed in Sect. 4.4 of the cited reference. In particular the dimensionality of the problem considered in the present paper ($d = 92$ with up to 53 dependent groups of features) is outside the scope of [6]'s algorithm.

The present work is mainly related to [16,17] insofar as it relies on a simple counting procedure on rectangular regions as in (4). As a comparison, [16,17] consider the truncated cones

$$\mathcal{C}_\alpha = \{x : \|x\|_\infty \geq 1, x_j > 0 \text{ for } j \in \alpha \; ; \; x_j = 0 \text{ for } j \notin \alpha\}. \tag{6}$$

The importance of such cones in the analysis comes from the homogeneity property of μ. More precisely, a subset of features α may take large values together while the others take small values, if and only if μ assigns a positive mass to $\mathcal{C}_\alpha$. The approach proposed in [17] consists in 'thickening' the cones $\mathcal{C}_\alpha$, i.e., defining for some small $\epsilon > 0$ (typically, $\epsilon = 0.1$),

$$\mathcal{C}_{\alpha,\epsilon} = \{x \in \mathbb{R}^d_+ : \|x\|_\infty \geq 1 \; ; \; \|x\|_\infty^{-1} x_j > \epsilon \text{ for } j \in \alpha \; ; \; \|x\|_\infty^{-1} x_j \leq \epsilon \text{ for } j \notin \alpha\}. \tag{7}$$

The quantity $\mu_\alpha := \mu(\mathcal{C}_\alpha)$ is approximated by its empirical counterpart on $\mathcal{C}_{\alpha,\epsilon}$, $\hat{\mu}(\mathcal{C}_\alpha) = \mu_n(\mathcal{C}_{\alpha,\epsilon})$, where μ_n is the empirical estimator defined in (5). In practice a tolerance parameter $\mu_{\min}$ has to be chosen: for any α such that $\mu_n(\mathcal{C}_{\alpha,\epsilon}) < \mu_{\min}$, one sets $\hat{\mu}(\mathcal{C}_\alpha) = 0$. The final output of [17]'s DAMEX algorithm is the potentially sparse $2^d - 1$-vector $\hat{\mathcal{M}} = (\hat{\mu}_\alpha)_{\alpha \subset \{1,\dots,d\}}$ mentioned in the introduction, with $\hat{\mu}_\alpha := \hat{\mu}(\mathcal{C}_\alpha)$.

One shortcoming of DAMEX is that no sparsity pattern is produced in case of 'noise'. Here, noise is understood as a small variability affecting the groups of features concomitantly impacted by an extreme event. As an example, for the hydrological dataset considered here, geophysics determines the main underlying dependence patterns, i.e., the groups of stations where floods *tend to* occur simultaneously (such as, say, group $\alpha_0 = \{1,2,3,4\}$); however due to meteorological variability, the actual observed floods sometimes affect some neighboring stations $5, 6$, so that in the dataset, the observed groups would be *e.g.* $\{\{1,2,3,4,5\}, \{1,2,3,4,6\}, \{1,2,3,4\}\}$. In such a case, the empirical mass is scattered over many sub-cones $\mathcal{C}_{\alpha,\epsilon}$ (three instead of one). This example suggests an alternative approach allowing to gather together those $\mathcal{C}_{\alpha,\epsilon}$'s that are 'close', as detailed next.

3.2 Gathering Together 'Close-By' Cones, Incremental Strategy

One way to gather different $\mathcal{C}_{\alpha,\epsilon}$'s together is to relax the condition that 'all the features V^j for $j \notin \alpha$ take small values' in the definition of $\mathcal{C}_{\alpha,\epsilon}$. This yields the rectangular region Γ_α defined in (4). Unlike the regions $\mathcal{C}_{\alpha,\epsilon}$'s, the Γ_α's do not form a partition of the positive orthant of $\mathbb{R}^d$, and indeed the fact that a point V_i belongs to Γ_α does not tell anything about its features V_i^j for $j \notin \alpha$. The problem addressed in [17] (recovering $\mathcal{M} := \{\alpha : \mu(\mathcal{C}_\alpha) > 0\}$) and the relaxed problem considered here (recovering $\mathbb{M} := \{\alpha : \rho_\alpha > 0\} = \{\alpha : \mu(\Gamma_\alpha) > 0\}$) are different but however related through the maximal elements of $\mathcal{M}$ and $\mathbb{M}$, as stated in the following lemma. Recall that α is said to be maximal in $\mathbb{M}$ (*resp.* $\mathcal{M}$) if there is no superset $\alpha' \supsetneq \alpha$ in $\mathbb{M}$ (*resp.* $\mathcal{M}$).

Lemma 1. *For $\alpha \subset \{1, \ldots, d\}$,*

$$\alpha \text{ is maximal in } \mathbb{M} \Leftrightarrow \alpha \text{ is maximal in } \mathcal{M}. \tag{8}$$

The proof is deferred to the Appendix.

Another important property from an algorithmic perspective is the following:

Lemma 2. *For $\alpha \subset \{1, \ldots, d\}$, if $\rho_\alpha = 0$ then also for all $\alpha' \supset \alpha$, $\rho_{\alpha'} = 0$.*

The proof is immediate: remind that $\rho_\alpha = \mu(\Gamma_\alpha)$ and notice that for $\alpha \subset \alpha'$, $\Gamma_{\alpha'} \subset \Gamma_\alpha$.

Apriori-Like Incremental Strategy. Lemma 2 suggests searching for α's satisfying $\rho_\alpha > 0$ following the Hasse diagram, among α's of increasing size, and stopping the search along a given path as soon as $\rho_\alpha = 0$ for some α. This incremental strategy is also the main ingredient of the Apriori algorithm ([2]), which we recall for convenience: Let $I = \{\text{item}_1, \ldots, \text{item}_d\}$ be set of items and let $T = \{t_1, \ldots, t_n\}$ be a set of transactions with $t_i \subset I, \forall i \in \{1, \ldots, n\}$. The frequency of occurrence of the list of items (itemset) $\alpha \subset I$ is defined as $f_\alpha := \frac{1}{n} \sum_{1 \leq i \leq n} \mathbb{1}_{\alpha \subset t_i}$. Apriori returns the set $\{\alpha : f_\alpha > f_{min}\}$ with $f_{min} > 0$. It begins with pairs of items and then increments the size of the itemsets at each step. Indeed if $f_\alpha \leq f_{min}$ then all supersets $\alpha' \supset \alpha$ verify $f_{\alpha'} \leq f_{min}$ as well, which reduces drastically the number of subsets to be tested.

The CLEF algorithm described next proceeds similarly: a concomitant occurrence of threshold excesses $\{V_i^j > t$ for features $j \in \alpha\}$ can be identified with a transaction and the dependence parameter ρ_α can be seen as a (rescaled) theoretical frequency. The main difference between CLEF and Apriori concerns the stopping criterion used by CLEF, which involves a *ratio* between frequencies for a group α and for subgroups $\beta \subset \alpha$. The idea behind is to allow detection of larger groups, as described in the following section.

4 Empirical Criterion and Implementation

4.1 Conditional Criterion for Extremal Dependence

Considering the relaxed framework where the goal is to recover the set $\mathbb{M}$ defined in (2), one needs an empirical criterion for testing the condition '$\rho_\alpha (= \mu(\Gamma_\alpha)) > 0$'. One option would be to consider the empirical estimator $\hat{\rho}_\alpha = \mu_n(\Gamma_\alpha)$ where

μ_n is defined in (5) which would be the (rescaled) counterpart of the empirical frequency f_α used in Apriori. Then the stopping criterion would be '$\hat{\rho}_\alpha \leq \rho_{\min}$', with $\rho_{\min}$ a user-defined tolerance level. However, since the Γ_α's (for increasing α's) are nested, the ρ_α's can only decrease with increasing sizes of α. In other words larger groups tend to be less frequent than smaller groups, even dependent ones. Thus in principle, detecting larger groups as well as smaller ones would require the tolerance level $\rho_{\min}$ to depend on the size $|\alpha|$ of the considered subgroup, which would result in $d - 1$ tuning parameters instead of one.

The alternative chosen in the present paper is to consider a *conditional* frequency, the conditioning event for a group α of size s being such that at least $s - 1$ features are large among the s considered ones. Now, there is no reason why the conditional frequency of occurrence should decrease with $|\alpha|$, so that a single tuning parameter needs to be chosen, without preventing the detection of large groups. In practice, computing conditional frequencies amounts to compare $\mu_n(\Gamma_\alpha)$ with $\mu_n(\Gamma_\beta)$, with $\beta \subset \alpha$. More precisely, let $\alpha \subset \{1, \ldots, d\}$ be such that for some $j \in \alpha$, $\rho_{\alpha \setminus \{j\}} > 0$. Consider the probability that all the features in α be large given that *all of them but at most one are large* and call κ_α the limiting conditional probability, namely

$$\kappa_\alpha = \lim_{t \to \infty} \frac{\mathbb{P}\left(\forall j \in \alpha, V_i^j > t\right)}{\mathbb{P}\left(\text{for all but at most one } j \in \alpha, V_i^j > t\right)}. \tag{9}$$

In the sequel, κ_α is referred to as the *conditional dependence coefficient* of α. Notice that the limit in (9) does exist: indeed, let

$$\Delta_\alpha = \cup_{j \in \alpha} \Gamma_{\alpha \setminus \{j\}} = \{x \in \mathbb{R}_+^d : \|x\|_\infty > 1, \sum_{j \in \alpha} \mathbb{1}_{x_j \geq 1} \geq |\alpha| - 1\}.$$

Since by assumption on α, for some j $\mu(\Gamma_{\alpha \setminus \{j\}}) = \rho_{\alpha \setminus \{j\}} > 0$, in view of (3), we have

$$\mu(\Delta_\alpha) = \lim_{t \to \infty} t\mathbb{P}(\text{ for all but at most one } j \in \alpha, V_i^j > t) > 0,$$

so that an equivalent definition of κ_α is

$$\kappa_\alpha = \frac{\lim_{t \to \infty} t\mathbb{P}\left(\forall j \in \alpha, V_i^j > t\right)}{\lim_{t \to \infty} t\mathbb{P}\left(\text{for all but at most one } j \in \alpha, V_i^j > t\right)}$$

$$= \frac{\mu(\Gamma_\alpha)}{\mu(\Delta_\alpha)}. \tag{10}$$

The idea is now to compare empirical counterparts of κ_α —using μ_n instead of μ, see (5)–with a single fixed tolerance parameter $\kappa_{\min} > 0$. This amounts to decide that $\mu_n(\Gamma_\alpha)$ results from noise if $\mu_n(\Gamma_\alpha) \ll \mu_n(\Delta_\alpha)$. Notice that $\Gamma_\alpha \subset \Delta_\alpha$, so that the empirical version of κ_α is again a conditional probability and thus belongs to $[0, 1]$ whenever $\mu_n(\Gamma_\beta) > 0$ for some $\beta \subset \alpha$ such that $|\alpha \setminus \beta| = 1$, which is another argument in favor of an incremental strategy.

4.2   Algorithm

CLEF (summarized in Algorithm 1) uses the empirical counterpart of the conditional criterion κ_α, which depends on a (high) threshold t as in (5):

$$\hat{\kappa}_{\alpha,t} := \frac{\mu_n(\Gamma_\alpha)}{\mu_n(\Delta_\alpha)} = \frac{\sum_{i=1}^n \mathbb{1}\{\#\{j \in \alpha: \hat{V}_i^j > t\} = |\alpha|\}}{\sum_{i=1}^n \mathbb{1}\{\#\{j \in \alpha: \hat{V}_i^j > t\} \geq |\alpha|-1\}}. \tag{11}$$

For $s \geq 2$, families $\hat{\mathcal{A}}_s$ of subsets α of size s are constructed in an incremental way, among a set of candidates $\mathcal{A}'_s$, as follows: Set $\hat{\mathcal{A}}_1 = \{\{1\}, \ldots, \{d\}\}$, then

$$\mathcal{A}'_s = \Big\{\alpha \subset \{1, \ldots, d\} : |\alpha| = s, \forall \beta \subset \alpha \text{ s.t. } |\beta| = s-1 : \beta \in \hat{\mathcal{A}}_{s-1}\Big\}$$

$$\hat{\mathcal{A}}_s = \Big\{\alpha \in \mathcal{A}'_s : \hat{\kappa}_{\alpha,t} > \kappa_{\min}\Big\}. \tag{12}$$

The procedure stops at step $S \leq d-1$ if $\hat{\mathcal{A}}_{S+1} = \emptyset$, at which point our estimator of the family $\mathbb{M}$ of dependent subsets is $\hat{\mathbb{M}} = \cup_{s=1}^S \hat{\mathcal{A}}_s$. Notice that restricting the search to the set of candidates $\mathcal{A}'_s$ ensures that the 'empirical counterpart' of Lemma 2 is satisfied, namely $\alpha \notin \hat{\mathbb{M}} \Rightarrow \forall \beta \supset \alpha, \beta \notin \hat{\mathbb{M}}$. It also avoids division by zero when computing (12). The final output of CLEF is the set $\hat{\mathbb{M}}_{\max}$ of maximal elements of $\hat{\mathbb{M}}$.

Remark 1 (Choice of the parameters t and $\kappa_{\min}$). The choice of t is a classical bias/variance trade-off: according to standard good practice in EVT (see e.g. [8]), t is chosen in a 'stability region' of relevant summaries of the output. Here we consider the cardinal of $\hat{\mathbb{M}}$ and the mean cardinal of maximal subsets $\alpha \in \hat{\mathbb{M}}$. When t is too small, the observed data may not have reached there ultimate regime (the extremal dependence structure characterized by μ in (3)), so that the bias of $\hat{\kappa}_{\alpha,t}$ may be large. In contrast, for too large values of t, very few excesses are observed so that the sample size of the data used to compute $\hat{\kappa}_{\alpha,t}$ is very small and the variance becomes too large. To wit, due to our standardization choice it holds that $\mathbb{P}(V_i^j > t) = 1/t$. Thus for each $j \in \{1, \ldots, d\}$, $|\{i : \hat{V}_i^j > t\}| \simeq 1/t$, so that the total number of data points for which at least one feature exceeds t is approximately within the interval $[n/t, dn/t]$. Results on real and simulated data (Sect. 5.3) bring out such a stability region for the above mentioned output summaries. It is empirically verified on simulated data that this region corresponds to near optimal values of t.

As for the tolerance parameter $\kappa_{\min}$, it should be chosen according to the context, keeping in mind that $\hat{\kappa}_{\alpha,t}$ is an empirical conditional probability of a joint threshold excess of all features $j \in \alpha$ (given that at least $|\alpha| - 1$ excesses have occurred). $\kappa_{\min}$ is the level above which this probability is considered as non negligible. The higher $\kappa_{\min}$, the more stringent the condition, the smaller and fewer the discovered groups α. In this work, we set $\kappa_{\min} = 0.25$.

Remark 2 (Construction of the candidates $\mathcal{A}'_{s+1}$). The graphical structure of the groups of features is exploited to construct candidate incremented groups

Algorithm 1. CLEF (CLustering Extreme Features)

INPUT: High threshold t, tolerance parameter $\kappa_{min} > 0$.

STAGE 1: constructing the $\hat{\mathcal{A}}_s$'s .

Initialization: set $S = d$.

Step 1: Construct the family of extremal-dependent pairs:
set $\hat{\mathcal{A}}_2 = \{\{i,j\} \subset \{1,\ldots,d\}, \text{ such that } \hat{\kappa}_{\{i,j\}} > \kappa_{min}\}$.

Step 2: If $\hat{\mathcal{A}}_2 = \emptyset$, set $S = 2$; end **STAGE 1**. Otherwise

- generate candidate triplets $\mathcal{A}'_3 = \{i,j,k\} \subset \{1,\ldots,d\}$ s.t $\{i,j\}, \{i,k\}, \{j,k\} \in \hat{\mathcal{A}}_2\}$,
- set $\hat{\mathcal{A}}_3 = \{\alpha \in \mathcal{A}'_3 \text{ s.t. } \hat{\kappa}_\alpha > \kappa_{min}\}$.

$\vdots$

Step $s(s \leq d)$: If $\hat{\mathcal{A}}_s = \emptyset$, set $S = s$; end **STAGE 1**. Otherwise

- generate candidates of size $s + 1$:
  $\mathcal{A}'_{s+1} = \{\alpha \subset \{1,\ldots,d\}, |\alpha| = s+1, \alpha \setminus \{j\} \in \hat{\mathcal{A}}_s \text{ for all } j \in \alpha\}$,
- set $\hat{\mathcal{A}}_{s+1} = \{\alpha \in \mathcal{A}'_{s+1} \text{ such that } \hat{\kappa}_\alpha > \kappa_{min}\}$.

Output: $\hat{\mathbb{M}} = \cup_{s=1}^{S} \hat{\mathcal{A}}_s$.

STAGE 2: pruning (keeping maximal α's only)

Initialization: $\hat{\mathbb{M}}_{max} \leftarrow \hat{\mathcal{A}}_S$.

for $s = (S-1) : 2$,   for $\alpha \in \hat{\mathcal{A}}_s$,

   If there is no $\beta \in \hat{\mathbb{M}}_{max}$ such that $\alpha \subset \beta$, $\hat{\mathbb{M}}_{max} \leftarrow \hat{\mathbb{M}}_{max} \cup \{\alpha\}$.

Output: $\hat{\mathbb{M}}_{max}$

---

of features. Namely, members of $\mathcal{A}'_{s+1}$ are the maximal cliques of size s in the graph $(\mathcal{A}_s, \mathcal{E}_s)$, where $\mathcal{E}_s = \{(\alpha, \alpha') \in \mathcal{A}_s \times \mathcal{A}_s : |\alpha \cap \alpha'| = s - 1\}$. The maximal clique problem is typically attacked *via* the max-clique algorithm ([29]). In the present work, clique extraction is performed using the function find_clique of the **Python** package **NetworkX**, which uses the Bron & Kerbosch ([5],[28]) algorithm.

5   Results

The aim of our experiments is threefold. First, CLEF's output on the hydrological data is illustrated and compared with DAMEX's (Sect. 5.1). Second, the respective performances of CLEF, DAMEX and Apriori are compared quantitatively on simulated data (Sect. 5.2). Finally (Sect. 5.3), the question of the threshold choice is investigated: the goal is to verify whether a stability region such as the one mentioned in Remark 1 exists and whether it corresponds to optimal performances of CLEF.

5.1   Stream-Flow Data

The output of CLEF for the stream-flow data may be visualized in Fig. 1 (Execution time: 0.09 s on a recent 4 cores laptop computer). Following the heuristic mentioned in Remark 1, the extremal threshold t was fixed to 320, yielding $k = 1186$ extreme events (time indexes i such that $\|\hat{V}_i\|_\infty \geq t$). The parameter $\kappa_{\min}$ was fixed to 0.25. A total number of 53 clusters (elements of $\hat{\mathbb{M}}_{\max}$) are returned by the CLEF algorithm, the size of which varies between 2 and 7. At first inspection, Fig. 1 agrees with general climatologic facts: in the north-western part of France, the climate is driven by large scale oceanographic perturbations, so that extreme floods tend to impact a large number of gauging stations simultaneously. The south-eastern part of France is rather subject to localized events (*e.g.* the so-called 'orages Cévenols' in the vicinity of the Mediterranean coast). This yields smaller clusters, both in terms of number of stations and of spatial extent.

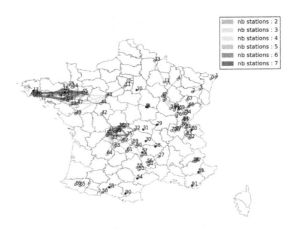

Fig. 1. Output of CLEF for the stream-flow dataset: maximal groups of stations $\alpha \in \hat{\mathbb{M}}$ that are likely to be jointly impacted by an extreme event. Clusters of stations are marked by colored edges between their members, the color scale indicates the number of stations forming the cluster. (Color figure online)

As a comparison, Table 1 shows the outcome of [17]'s DAMEX algorithm with the stream-flow data. These results show that no matter the choice of the thickening parameter ϵ in (7), the data do not concentrate on 'a few' thickened cones $\mathcal{C}_{\alpha,\epsilon}$, instead most of the empirical mass is spread onto many of them. In other words, there are too many subcones with positive mass, but not in a significant way.

5.2   Simulation Experiments

In order to quantify the relative performances of CLEF, DAMEX and Apriori, we generate d-dimensional datasets under a model such that the exponent measure μ concentrates on p specified cones $(\mathcal{C}_{\alpha_1}, \ldots, \mathcal{C}_{\alpha_p})$. Notice that $p, (\alpha_1, \ldots, \alpha_p)$ only determine the generating model, they are not used as inputs of any of the three algorithms compared here. The generated data are 'realistic' in the sense that all

Table 1. Output of [17]'s DAMEX algorithm with the hydrological dataset. Columns 1 and 2 indicate respectively the number of thickened cones $\mathcal{C}_{\alpha,\epsilon}$ with non zero empirical mass, and the percentage of cones (among those such that $\mu_n(\mathcal{C}_{\alpha,\epsilon}) > 0$) containing less than 1% of the 'extreme data', that is of $\#\{i : \|\hat{V}_i\|_\infty > t\}$.

ϵ	$\#\left\{\alpha : \mu_n(\mathcal{C}_{\alpha,\epsilon}) > 0\right\}$	$\%\left\{\alpha : \frac{\#\{i:t^{-1}V_i \in \mathcal{C}_{\alpha,\epsilon}\}}{\#\{i:\|V_i\| \geq t\}} < 1\%\right\}$
0.01	740	100%
0.05	688	98%
0.1	639	94%
0.2	559	88%

the features are positive (the points lie in the interior cone $\mathcal{C}_{\{1,\dots,d\}}$), even though the furthest points in the tails concentrate near the subcones $\mathcal{C}_{\alpha_k}$'s. Namely, we use the *asymmetric logistic* extreme value model ([27]), from which data is simulated using Algorithm 2.2 in [26]. 20 datasets of size $n = 100 \cdot 10^3, d = 100$, are generated. For each dataset, p subsets $\alpha_1, \dots, \alpha_p$ of $\{1, \dots, d\}$ are randomly chosen, which sizes follow a truncated geometric distribution (the maximum cluster size is 8). We aim at reproducing the fact that different events associated with a single α usually impact a group of stations which differs from α by a few stations (the impacted area is not deterministic). To this end, for each step $i = 1, \dots, n$, and each subset α_j, $j = 1, \dots, p$, one randomly chosen 'noisy' feature $l_{i,j} \in \{1, \dots, d\} \setminus \alpha_j$ is added to α_j. For CLEF, DAMEX and Apriori algorithms, the extreme threshold parameter t is chosen so that $\frac{\#\{i \leq n : \|\hat{V}_i\|_\infty \geq t\}}{n} \approx 5\%$. Table 2 summarizes the average performance of the three algorithms, for $p = 40, 50, 60, 70$. In these experiments, the CLEF algorithm recovers most of the charged p subsets $\alpha_1, \dots, \alpha_p$ in average, and significantly more than Apriori. In contrast, DAMEX does not recover the sparse structure of the data. It should be noted that in situations (not reported here) where no noisy feature is added, Apriori and DAMEX perform as well as CLEF.

Table 2. Average number of errors (non recovered and falsely discovered clusters) of CLEF, Apriori and DAMEX with simulated, noisy data.

p	# errors CLEF	# errors Apriori	# errors DAMEX
40	1.2	6.4	72.2
50	3.5	10.9	91.0
60	6.3	14.6	112.4
70	10.1	25.8	134.0

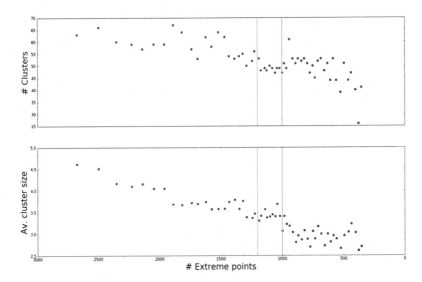

Fig. 2. Stability region for k (number of extreme points) on the stream-flow data. Upper panel: number of detected clusters, lower panel: average cluster size. Vertical red lines ($k \in \{1000, \ldots, 1200\}$ / $t \in [320, 400]$): stability region. (Color figure online)

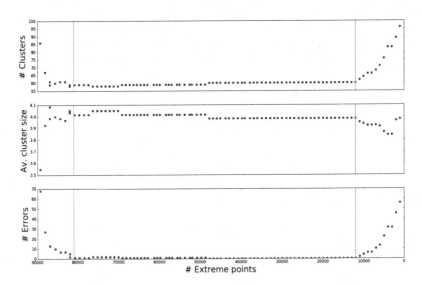

Fig. 3. Stability region for k (number of extreme points) on simulated data. Upper panel: number of detected clusters, middle panel: average cluster size, lower panel: number of errors of CLEF (as in Table 2). Vertical red lines ($k \in \{12000, \ldots, 81000\}$): stability region. (Color figure online)

5.3   Influence of the Threshold Choice

The high threshold t plays a decisive role in our framework as it determines which standardized features V_i^j are considered as extreme. Recall that the estimate $\hat{V}_i^j$ is discrete (see Sect. 2.2). A more convenient way to evaluate the influence of the threshold t is thus to consider instead $k := \#\{i \in \{1, \ldots, n\} : \|\hat{V}_i\|_\infty > t\}$ the total number of extreme points. Two significant summaries of CLEF output which are the number of clusters $|\hat{\mathbb{M}}|$ and their average sizes $\frac{1}{|\hat{\mathbb{M}}|} \sum_{\alpha \in \hat{\mathbb{M}}} |\alpha|$, are plotted as a function of k. Figure 2 (hydrological data) and the first two panels of Fig. 3 (simulated data) confirm the existence of stability regions (vertical red lines). The simulation experiments show that choosing the parameter in such regions ensures an optimal performance for CLEF, since both match exactly the one of lowest errors. The large width of the stability region for the simulated data (Fig. 3 compared to Fig. 2) may be explained by the fact that the generative model is a classical parametric extreme value model for which the asymptotic regime is nearly reached even for small thresholds, leading to large stability regions.

6   Conclusion

We propose a novel dimension reduction method for the analysis of extremes of multivariate datasets *via* feature clustering. This is done in adequacy with the framework of multivariate extreme value theory. The proposed algorithm makes use of the graphical structure of the problem, scanning the multiple possible subsets of features in a time efficient way. Results on a hydrological stream-flow data and on simulated data demonstrate the relevance of this approach on datasets which would not exhibit any sufficiently sparse structure when analyzed with existing algorithms. Future work will focus on the statistical properties of the empirical criteria $\hat{\kappa}_{\alpha,t}$ involved in the algorithm, which would allow to analyze the output as a statistical test for independence at extreme levels.

Acknowledgments. Part of this work has been funded by the the 'LabEx Mathématiques Hadamard' (LMH) project, by the 'AGREED' project from the PEPS JCJC program (INS2I, CNRS) and by the chair 'Machine Learning for Big Data' from Télécom ParisTech. The authors would like to thank Benjamin Renard for interesting discussions about the hydrological use case and for sharing the data.

A   Appendix: Proof of Lemma 1

Step 1. As a first step we show that $\mathcal{M} \subset \mathbb{M}$, i.e., $\mu(\mathcal{C}_\alpha) > 0 \Rightarrow \mu(\Gamma_\alpha) > 0$.

Proof Write $\mathcal{C}_\alpha = \bigcup_{\epsilon > 0, \epsilon \in \mathbb{Q}} R_{\alpha,\epsilon}$, where $R_{\alpha,\epsilon} = \{x \in \mathbb{R}_+^d : \|x\|_\infty \geq 1;\ x_j > \epsilon\ (j \in \alpha);\ x_i = 0\ (i \notin \alpha)\}$. Assume $\mu(\mathcal{C}_\alpha) > 0$. Since $\mu(\mathcal{C}_\alpha) < \infty$, by the monotonous limit property of the measure μ, we have $\mu(\mathcal{C}_\alpha) = \lim_{\epsilon \to 0} \mu(R_{\alpha,\epsilon})$. Also, from the definitions, $R_{\alpha,\epsilon} \subset \epsilon\Gamma_\alpha$. Thus,

$$\mu(\mathcal{C}_\alpha) > 0 \Rightarrow \exists \epsilon \in (0,1) : \mu(R_{\alpha,\epsilon}) > 0 \qquad \Rightarrow \mu(\epsilon\Gamma_\alpha) > 0$$
$$\Rightarrow \rho_\alpha = \mu(\Gamma_\alpha) = \epsilon\mu(\epsilon\Gamma_\alpha) > 0.$$

Step 2. We now prove the reverse inclusion for maximal elements of $\mathbb{M}$, i.e.,

$$\alpha \text{ is maximal in } \mathbb{M} \quad \Rightarrow \alpha \in \mathcal{M}. \tag{13}$$

Proof Consider, for $i \notin \alpha$, the set $\Delta_{i,\epsilon} = \Gamma_\alpha \cap \{x \in \mathbb{R}_+^d : \quad x_i > \epsilon\}$, so that $\Gamma_\alpha = \left\{ \bigcup_{\substack{i \in \{1,\ldots,d\} \setminus \alpha \\ \epsilon \in \mathbb{Q} \cap (0,1)}} \Delta_{i,\epsilon} \right\} \cup R_{\alpha,1}$. Thus,

$$\alpha \in \mathbb{M} \quad \Rightarrow \quad \mu(\Gamma_\alpha) > 0 \quad \Rightarrow \quad \left(\exists i, \mu(\Delta_{i,\epsilon}) > 0 \quad \text{or} \quad \mu(R_{\alpha,1}) > 0 \right) \tag{14}$$

To prove (13), it is enough to show that

$$\alpha \in \mathbb{M} \quad \Rightarrow \quad \text{for } i \notin \alpha, \ \mu(\Delta_{i,\epsilon}) = 0. \tag{15}$$

Indeed if (15) is true, and if $\alpha \in \mathbb{M}$, then (14) implies that $\mu(R_{\alpha,1}) > 0$, and the result follows from the inclusion $R_{\alpha,1} \subset \mathcal{C}_\alpha$. We show (15) by contradiction. If $\mu(\Delta_{i,\epsilon}) > 0$ for some $i \notin \alpha$, then

$$\frac{1}{\epsilon} \Delta_{i,\epsilon} = \left(\frac{1}{\epsilon} \Gamma_\alpha \right) \cap \{x \in \mathbb{R}_+^d : x_i > 1\} \subset \Gamma_{\alpha \cup \{i\}},$$

thus $\mu(\Gamma_{\alpha \cup \{i\}}) > 0$, which contradicts the maximality of α in $\mathbb{M}$.

Step 3. From (13), if α is maximal in $\mathbb{M}$ then $\alpha \in \mathcal{M}$. Now if α is maximal in $\mathbb{M}$ but not in $\mathcal{M}$, there exists $\beta \supsetneq \alpha$ in $\mathcal{M}$. Thus from Step 1, $\beta \in \mathbb{M}$, a contradiction. Hence α is also maximal in $\mathcal{M}$. Conversely, if α is maximal in $\mathcal{M}$ then (Step 1) $\alpha \in \mathbb{M}$. If α was not maximal in $\mathbb{M}$, there would exist $\beta \supsetneq \alpha$ maximal in $\mathbb{M}$, and from (13), $\beta \in \mathcal{M}$, contradicting the maximality of α in $\mathcal{M}$.

References

1. Agrawal, R., Gehrke, J., Gunopulos, D., Raghavan, P.: Automatic subspace clustering of high dimensional data. DMKD **11**(1), 5–33 (2005)
2. Agrawal, R., Srikant, R., et al.: Fast algorithms for mining association rules. In: Proceedings of 20th International Conference on Very Large Data Bases, VLDB, vol. 1215, pp. 487–499 (1994)
3. Beirlant, J., Goegebeur, Y., Segers, J., Teugels, J.: Statistics of extremes: theory and applications. Wiley, Hoboken (2006)
4. Boldi, M.O., Davison, A.: A mixture model for multivariate extremes. JRSS-B **69**(2), 217–229 (2007)
5. Bron, C., Kerbosch, J.: Algorithm 457: finding all cliques of an undirected graph. Commun. ACM **16**(9), 575–577 (1973)
6. Chautru, E.: Dimension reduction in multivariate extreme value analysis. Electron. J. Stat. **9**(1), 383–418 (2015)
7. Clifton, D.A., Hugueny, S., Tarassenko, L.: Novelty detection with multivariate extreme value statistics. J. Sig. Process. Syst. **65**(3), 371–389 (2011)
8. Coles, S.: An Introduction to Statistical Modeling of Extreme Values. Springer Series in Statistics. Springer, London (2001)

9. Coles, S., Tawn, J.: Modeling extreme multivariate events. JRSS-B **53**, 377–392 (1991)
10. Cooley, D., Davis, R., Naveau, P.: The pairwise beta distribution: a flexible parametric multivariate model for extremes. JMVA **101**(9), 2103–2117 (2010)
11. Einmahl, J.H., Segers, J.: Maximum empirical likelihood estimation of the spectral measure of an extreme-value distribution. Ann. Stat. **37**, 2953–2989 (2009)
12. Fougeres, A.L., De Haan, L., Mercadier, C., et al.: Bias correction in multivariate extremes. Ann. Stat. **43**(2), 903–934 (2015)
13. Fougeres, A.L., Mercadier, C., Nolan, J.P.: Dense classes of multivariate extreme value distributions. J. Multivar. Anal. **116**, 109–129 (2013)
14. Giuntoli, I., Renard, B., Vidal, J.P., Bard, A.: Low flows in france and their relationship to large-scale climate indices. J. Hydro. **482**, 105–118 (2013)
15. Goix, N., Sabourin, A., Clémençon, S.: Learning the dependence structure of rare events: a non-asymptotic study. In: Proceedings of the 28th COLT (2015)
16. Goix, N., Sabourin, A., Clémençon, S.: Sparsity in multivariate extremes with applications to anomaly detection. arXiv preprint arXiv:1507.05899 (2015)
17. Goix, N., Sabourin, A., Clémençon, S.: Sparse representation of multivariate extremes with applications to anomaly ranking. In: Proceedings of the 19th AISTAT conference, pp. 287–295 (2016)
18. Guillotte, S., Perron, F., Segers, J.: Non-parametric Bayesian inference on bivariate extremes. JRSS-B **73**(3), 377–406 (2011)
19. Gunopulos, D., Khardon, R., Mannila, H., Saluja, S., Toivonen, H., Sharma, R.S.: Discovering all most specific sentences. ACM Trans. Database Syst. **28**(2), 140–174 (2003)
20. Katz, R.W., Parlange, M.B., Naveau, P.: Statistics of extremes in hydrology. Adv. Water Resour. **25**(8), 1287–1304 (2002)
21. Lee, H.-J., Roberts, S.J.: On-line novelty detection using the Kalman filter and extreme value theory. In: 19th International Conference on Pattern Recognition, ICPR 2008, pp. 1–4. IEEE (2008)
22. Qi, Y.: Almost sure convergence of the stable tail empirical dependence function in multivariate extreme statistics. Acta Math. Applicatae Sin. (English Ser.) **13**(2), 167–175 (1997)
23. Resnick, S.I.: Extreme Values, Regular Variation and Point Processes. Springer, Heidelberg (2013)
24. Sabourin, A., Naveau, P.: Bayesian Dirichlet mixture model for multivariate extremes: a re-parametrization. CSDA **71**, 542–567 (2014)
25. Sabourin, A., Naveau, P., Fougeres, A.L.: Bayesian model averaging for multivariate extremes. Extremes **16**(3), 325 (2013)
26. Stephenson, A.: Simulating multivariate extreme value distributions of logistic type. Extremes **6**(1), 49–59 (2003)
27. Tawn, J.A.: Modelling multivariate extreme value distributions. Biometrika **77**(2), 245–253 (1990)
28. Tomita, E., Tanaka, A., Takahashi, H.: The worst-case time complexity for generating all maximal cliques and computational experiments. Theoret. Comput. Sci. **363**(1), 28–42 (2006)
29. Xie, Y., Philip, S.Y.: Max-clique: a top-down graph-based approach to frequent pattern mining. In: 2010 IEEE International Conference Data Mining, pp. 1139–1144. IEEE (2010)

Profiling Human Behavior Through Multidimensional Latent Factor Modeling

Massimo Guarascio[(✉)], Francesco Sergio Pisani, Ettore Ritacco,
and Pietro Sabatino

Institute for High Performance Computing and Networking of the Italian National
Research Council, ICAR - CNR, New Delhi, India
massimo.guarascio@icar.cnr.it

Abstract. The Human Behavioral Analysis is a growing research area
due to its big impact on several scientific and industrial applications. One
of the most popular family of techniques addressing this problem is the
Latent Factor Modeling which aims at identifying interesting features
that determine human behavior. In most cases, latent factors are used
to relate atomic features to each other: for example, semantically similar
words in documents of a textual corpus (text analysis), products to buy
and customers (recommendation), users (social influence) or news in a
social network (information diffusion). In this paper, we propose a new
latent-factor-based approach whose goal is to profile users according to
their behavior. The novelty of our proposal consists in considering the
actions as set of features instead of single atomic elements. A single
action is characterized by several components that can be exploited in
order to define fine-grain user profiles. These components can be, for
instance, "what is being done", "where", "when" or "how". We evaluated
our approach in two application scenarios. A first test is performed on
real data and it is aimed at semantically validate the model identifying
behavioral clusters of users; a second test is a predictive experiment on
synthetic data generated to assess model's anomaly detection capability.

1 Introduction

Increasing attention has been paid to the problem of identifying and explaining
human activities based on the user behavior. The concept of behavior is a relevant aspect for several challenging application domains such as Recommender
Systems, Cyber Security, Fraud Detection, Surveillance and Fault Detection Systems. The aim is to learn the usual/normal behavior of a person in order to better
understand how she acts and to detect potential anomalies.

Behavior Computing [6] is an emerging research field whose goal is to investigate mathematical models that could summarize the dynamics of a complex system such as a human being. Several big companies such as Facebook[1], Twitter[2],

---

[1] https://www.facebook.com/.

[2] https://twitter.com/.

© Springer International Publishing AG 2017
A. Appice et al. (Eds.): NFMCP 2016, LNAI 10312, pp. 148–162, 2017.
DOI: 10.1007/978-3-319-61461-8_10

Youtube [3] or Tumblr[4] are investing resources and money to equip their services with advanced analytic functions able to discover user profiles. These profiles are used to learn the causes that triggers users' actions, with the aim to recommend items that match their taste, suggest personalized information, share interesting contents or propose social connections, detect anomalous events/actions (e.g. due to identity theft).

Several research areas are addressing the Behavior Computing problem under different perspectives. Particular interest has been shown for Latent Factor Modeling [16] that is able to define and estimate a set of unobserved variables, called latent factors or topics, summarizing the observable features of an underlying data sample. Considering as sample a human activity log, data are the set of the actions per user and the latent factors are the unobserved causes that explain somehow why the user performed those actions. Hence, a distribution probability over the latent factors can be translated as a behavior profile for the target user.

Literature is rich of latent factor techniques aimed to model complex behaviors. Some example are. [1,4,7,11,13]. All of these approaches share the assumption that actions are represented as atomic data or sometimes as a twofold piece of information: what is done and when. In other words, most part of the Behavior Computing approaches assumes that the underlying phenomenon is characterized by elementary elements governed by some distribution probability, for instance:

- Recommender systems deal with users and items (in many cases as *ID*);
- Link predictors in social networks deal with users and their atomic features;
- Community detectors deal with humans and links;
- Information diffusion predictors deal with humans and topics.

In this work we propose a novel technique for discovering user profiles able to handle multidimensional data which is based on an extended version of Latent Dirichlet Allocation (LDA) [5]. As aforesaid, human behavior is defined according to the set of actions performed by a person, but each action is multivariate since it is composed by a set of features: what is done, where and when, how, near what, and so on represent some types of these features. In other words, actions are not atomic. The contribution to the literature of the proposed approach consists in the definition a fine-grained human behavior learning schema where each user action is analyzed from different angles of view (contexts or dimensions). The objective is to capture a full set of aspects that characterize human activities. For a better understanding, consider the action "a user is entering in a bank", there is a big difference if the action is performed during the day or the night.

In literature there is a strong research field that deals with multivariate latent factor models, such as the Gaussian Processes. However, the main drawback of this kind of models is the computational complexity of a prediction: for instance,

---

[3] https://www.youtube.com/.
[4] https://www.tumblr.com/.

in many definition of gaussian process, the complexity is cubic in the number of elements [2,17], hence, it does not scale for very large dataset.

The rest of the paper is organized as follows. Section 2 formally defines the proposed approach and explains how perform the parameter estimation; Sect. 3 describes the multivariate nature of the data which will be used to learn the model; Sect. 4 evaluates the model in two application scenarios; finally, Sect. 5 concludes the paper.

2   Multidimensional Latent Dirichlet Allocation

In this paper we propose a model based on Latent Dirichlet Allocation (LDA) [5], and basically we extend this technique to a multidimensional setting. Indeed, LDA usually considers a dyadic system and then tries to establish an association scheme between two related sets of atomic elements, for instance: documents and words, users and items, etc. Observing that there are scenarios in which at least one of the two considered sets may actually contain composed elements, i.e. that are characterized by multiple values that range in domains of different dimensions, it is then natural to try to employ LDA related techniques in this more general setting. The proposed model is then named Multidimensional Latent Dirichlet Allocation, (MDLDA); for reader convenience, in Table 1 we summarize all the variables and quantities used in the model definition, Fig. 1 gives a graphical overview of the model and, finally, its generative process is given in Table 2.

An action a performed by a user u is composed by a set of observable values $\{i^{(1)}, \ldots, i^{(n)}\}$ belonging to n different domains, one for each dimension characterizing a. We assume that each user u performs A_u actions, each one is governed by a latent variable z that represents a topic. Given a topic $z = k$ where $k \in \{1, \ldots, K\}$, the observable values are sampled according the *Discrete* distributions $\phi_k^{(d)}$, with $d \in \{1, \ldots, n\}$. The value of the latent variable z is sampled from a *Discrete* distribution dependent on u, θ_u, we are assuming that our data set is composed by the actions of U users. Each *Discrete* distribution is generated by *Dirichlet* distributions, whose hyper parameters are α for θ_u and β^d for each $\phi_k^{(d)}$, where $d \in \{1, \ldots, n\}$ and $k \in \{1, \ldots, K\}$.

2.1   Parameter Inference

In what follows we are going to infer parameters for the proposed model, MDLDA. First of all, let us consider the complete-data likelihood:

$$\Pr(\mathbf{A}, \mathbf{Z}, \mathbf{\Theta}, \mathbf{\Phi} | \alpha, \beta) =$$
$$\left\{ \prod_{d=1}^{n} \prod_{k=1}^{K} \Pr\left(\phi_k^{(d)} | \beta^{(d)}\right) \right\} \prod_{u \in U} \Pr(\theta_u | \alpha) \prod_{a \in \mathbf{A}_u} \Pr(z_{u,a} | \theta_u) \prod_{d \in D} \Pr\left(i_{u,a}^{(d)} | \phi_{z_{u,a}}^{(d)}\right),$$

Table 1. Notation table.

Variable	Description
U	The set of users u
K	Number of topics
Z	Latent variable
D	The set of all dimensions, the index of a dimension is d, and $|D| = n$
A	The set of actions, A_u is the actions taken by user u
$\vec{\alpha}$	Hyperparameter mixture on users (a vector of K elements or a scalar if symmetric)
$\vec{\beta}^{(d)}$	Hyperparameter mixture on topics for each dimension d (a vector of V elements or a scalar if symmetric)
θ_u	Profile of user u, i.e. multinomial distribution over number of topics K
$\phi_k^{(d)}$	Mixture components over the elements given a topic k and a dimension d
$\left(i_a^{(1)}, \ldots, i_a^{(n)}\right)$	Vector representing action a where each $i_a^{(d)}$ is the value of dimension d within action a. Given Z, values in each dimension are conditionally independent from each other

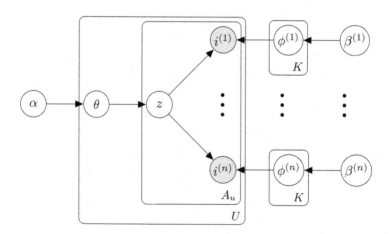

Fig. 1. Multidimensional Latent Dirichlet Model in plate notation.

by applying to θ and ϕ the Law of Total Probability, from the above equation we obtain a formula for the complete data likelihood:

$$\Pr\left(\mathbf{A}, \mathbf{Z} | \alpha, \beta\right) = \int_\theta \int_{\phi^{(1)}} \ldots \int_{\phi^{(n)}} \Pr\left(\mathbf{A}, \mathbf{Z}, \mathbf{\Theta}, \mathbf{\Phi} | \alpha, \beta\right) d\theta d\phi^{(1)} \ldots d\phi^{(n)}.$$

Since the integrals in the above equality are independent, we can group them in the following way:

$$\left(\int_\theta \prod_{u \in U} \Pr\left(\theta_u | \alpha\right) \prod_{a \in \mathbf{A}_u} \Pr\left(z_{u,a} | \theta_u\right) d\theta \right) \cdot$$

$$\left(\prod_{d=1}^{n} \int_{\phi^{(d)}} \left[\prod_{k=1}^{K} \Pr\left(\phi_k^{(d)} | \beta^{(d)}\right) \right] \prod_{u \in U} \prod_{a \in \mathbf{A}_u} \Pr\left(i_{u,a}^{(d)} | \phi_{z_{u,a}}^{(d)}\right) d\phi^{(d)} \right). \tag{1}$$

Table 2. Generative process for the proposed model

- For every user $u \in U$ choose $\theta_u \sim Dirichlet\,(\alpha)$
- For every dimension $d \in D$
  - Choose $\phi_k^{(d)} \sim Dirichlet\left(\beta^{(d)}\right)$ for $k \in \{1, \ldots, K\}$
- For every user $u \in U$ and every action $a \in A_u$
  - Choose the latent variable $z \sim Discrete\,(\theta_u)$
  - For every dimension $d \in D$
    * Choose a value $i^{(d)} \sim Discrete\left(\phi_z^{(d)}\right)$

In the above formula, let us consider the first integral, after introducing some notation, we are going to reformulate it. First of all, denote by n_u^k the number of times a user u get associated to the latent variable k, we then have:

$$\prod_{a \in \mathbf{A}_u} \Pr\left(z_{u,a} | \theta_u\right) = \prod_{k=1}^{K} \theta_{u,k}^{n_u^k},$$

since $\Pr\left(\theta_u | \alpha\right)$ is given by a Dirichlet distribution, then

$$\Pr\left(\theta_u | \alpha\right) = \frac{1}{\Delta\,(\alpha)} \prod_{k=1}^{K} \theta_{u,k}^{\alpha_k - 1},$$

where for every vector $\mathbf{w} = \{w_1, \ldots, w_r\}$, $\Delta\,(\mathbf{w})$ is given by:

$$\Delta\,(\mathbf{w}) = \frac{\prod_{i=1}^{r} \Gamma\,(w_i)}{\Gamma\left(\sum_{i=1}^{r} w_i\right)},$$

and as usual by $\Gamma\,(\cdot)$ we denote the Gamma function. By a straightforward computation we can then rewrite the first integral in (1) as:

$$\int_\theta \prod_{u \in U} \Pr\left(\theta_u | \alpha\right) \prod_{a \in \mathbf{A}_u} \Pr\left(z_{u,a} | \theta_u\right) d\theta = \prod_{u \in U} \frac{\Delta\,(\alpha + \mathbf{n}_u)}{\Delta\,(\alpha)},$$

where $\mathbf{n}_u = \{n_u^k\}_{k=1}^K$. By an analogous derivation, we are able to reformulate the second integral in (1). Indeed we have:

$$\Pr\left(\phi_k^{(d)}|\beta^{(d)}\right) = \frac{1}{\Delta(\beta)} \left(\prod_{i^{(d)} \in val(d)} \left(\phi_{k,i^{(d)}}^{(d)}\right)^{\beta_{i^{(d)}}-1} \right),$$

and

$$\prod_{u \in U} \prod_{a \in \mathbf{A}_u} \Pr\left(i_{u,a}^{(d)}|\phi_{z_{u,a}}^{(d)}\right) = \prod_{k=1}^K \prod_{i^{(d)} \in val(d)} \left(\phi_{k,i^{(d)}}^{(d)}\right)^{n_{i^{(d)}}^k},$$

then

$$\prod_{d=1}^n \int_{\phi^{(d)}} \left[\prod_{k=1}^K \Pr\left(\phi_k^{(d)}|\beta^{(d)}\right) \right] \prod_{u \in U} \prod_{a \in \mathbf{A}_u} \Pr\left(i_{u,a}^{(d)}|\phi_{z_{u,a}}^{(d)}\right) d\phi^{(d)} = \prod_{d=1}^n \prod_{k=1}^K \frac{\Delta\left(\beta+\mathbf{n}_d^k\right)}{\Delta(\beta)},$$

where $\mathbf{n}_d^k = \{n_{i^{(d)}}^k\}_{i^{(d)} \in val(d)}$, $n_{i^{(d)}}^k$ is a counter of how many times a topic k is related in dimension d with the item $i^{(d)}$ and $val(d)$ is the set of all possible values of the dimension d. Finally, we are able to write the complete-data likelihood:

$$\Pr\left(\mathbf{A}, \mathbf{Z}|\alpha, \beta\right) = \left(\prod_{u \in U} \frac{\Delta(\alpha + \mathbf{n}_u)}{\Delta(\alpha)} \right) \cdot \prod_{d=1}^n \prod_{k=1}^K \frac{\Delta\left(\beta+\mathbf{n}_d^k\right)}{\Delta(\beta)}. \tag{2}$$

2.2   Final Equations

It is clear that the formulation of the likelihood in Eq. (2) is rather hard to manage from a computational point of view, and as usual, we are forced to use a simpler approximation of it by some sort of heuristic argument. In order to achieve this goal, in the present work, we choose Gibbs Sampling technique, see for instance [9,10,15], that belongs to the family of Monte Carlo methods. At this point we are able to summarize the MDLDA model that we are going to employ in our experiments.

- Sampling of latent variables:

$$\Pr\left(z_{u,a} = k|Z_{-u,a}, A; \alpha, \beta\right) \propto$$
$$\left(n_u^k + \alpha_k - 1\right) \prod_{d \in Dim(a)} \frac{n_{i^{(d)}}^k + \beta_{i^{(d)}}^{(d)} - 1}{-1 + \sum_{j^{(d)}} n_{j^{(d)}}^k + \beta_{j^{(d)}}^{(d)}}. \tag{3}$$

- Computation of behavioral profiles:

$$\theta_{u,k} = \frac{n_u^k + \alpha_k}{\sum_{h=1}^K n_u^h + \alpha_h}. \tag{4}$$

– Computation of similarity between domains and latent variables:

$$\phi^d_{k,i(d)} = \frac{n^k_{i(d)} + \beta^{(d)}_{i(d)}}{\sum_{j(d)} n^k_{j(d)} + \beta^{(d)}_{j(d)}}. \tag{5}$$

– Moreover we add a module that updates hyper parameters of behavioral profiles:

$$\alpha^{new}_k = \alpha_k \frac{\sum_{u \in U} \left[\Psi\left(n^k_u + \alpha_k\right) - \Psi\left(\alpha_k\right) \right]}{\sum_{u \in U} \left[\Psi\left(\sum_{k=1}^{K} n^k_u + \alpha_k\right) - \Psi\left(\sum_{k=1}^{K} \alpha_k\right) \right]}, \tag{6}$$

where $\Psi\left(\cdot\right)$ represents the function Digamma, see Minka [14] for the above formulation.

2.3   Inference Algorithm

We are now ready to present our algorithm for inferring behavioral profiles in the MLDA model, it is based on the formulation contained in the preceding section, and it is illustrated in the pseudocode contained in Algorithm 1.

```
input  : nIterations, maximal number of iterations; burnIn, number of iterations of the burn-in step
output:  Θ behavioral profiles matrix; Φ^(d) latent variables – values of dimensions matrix
1   Θ* ← ∅
2   for d ∈ D do
3   |     Φ*^(d) ← ∅
4   end
5   for u ∈ U do
6   |     for a ∈ A_u do
7   |     |     Choose latent factor z_{u,a} ~ Uniform(K)
8   |     end
9   end
10  for it ← 1 to nIterations do
11  |     for u ∈ U do
12  |     |     for a ∈ A_u do
13  |     |     |     Update z_{u,a} ~ Pr(z_{u,a}|Z_{-u,a}, A; α, β), (3)
14  |     |     end
15  |     end
16  |     if it > burnIn ∧ (it ≡ 0 mod sampleLag) then
17  |     |     Compute Θ, (4)
18  |     |     Θ* = Θ* ∪ Θ
19  |     |     for d ∈ D do
20  |     |     |     Compute Φ^(d), (5)
21  |     |     |     Φ*^(d) = Φ*^(d) ∪ Φ^(d)
22  |     |     end
23  |     end
24  |     Update α, (6)
25  end
26  return mean(Θ*) and for each d ∈ D, mean(Φ*^(d))
```

Algorithm 1: Pseudocode for MDLDA's parameter inference.

The algorithm starts with a random initialization of latent variables involved and proceeds with a fixed point search procedure based on Gibbs Sampling. This fixed point search comprises two main steps. The first one, that we call *burn-in*, intended as a measure to mitigate the influence of the random initialization, resamples various times latent variables. As soon as this burn-in phase is finished, in the subsequent step the algorithm performs various iterations of the *sample-lag* procedure and in each iteration user profiles are computed. Once the maximal number of iterations are reached, the algorithm returns users behavioral profiles as mean over all results of sample-lag iterations, as well as associations between latent variables and contextual domains values.

3   Data Model

The model-feeding data are based on three main concepts: users, actions and dimensions. In particular, in our approach, a user is represented by a numeric identifier, actions are n-ary vectors of discrete values which are defined on domains that represent the various dimensions. A dimension is a suitable feature able to define an aspect of the action, for instance it may be the object of the action, a time stamp that keeps track of when the action has been performed, where the action took place, number and types of devices involved (e.g. computer, telephone, etc.), files that are accessed during the action, active services and so on. Hence, an action is defined by the set of its dimension values and the sequences of the user actions are processed in order to discover typical behavioral profiles within a user set.

3.1   Data Manipulation

First of all, it is worth noting that the dimensions, specifically their number and type, depend on the scenario to be analyzed. For instance in a scenario related to a business environment, some dimensions are related to security and permissions issues. They are useful to describe a general user behavior, but in a scenario where all users are granted the same rights or share unrestricted access to a set of places/documents those types of dimensions are no more relevant.

In our proposal we assume that each dimension domain is discrete and finite, i.e. it a characterizing feature with a fixed number of possible values. In this process then, each real value (e.g. the time for a time context, the places for a location type, etc.) needs to be converted to a nominal value in a previous pre-processing phase.

According to this definition, the target data structure is a sparse three-dimensional tensor: the first dimension represents users, the second one actions, while the last one the action features. Each user action is defined as a tuple whose fields are the user unique identifier, the action unique identifier and a sequence of integers mapping all dimension values.

As stated above, data are stored in a cubic structure in which the users, the performed actions and the dimensions are, respectively, represented by the first, second and third cube's dimension. A possible software implementation of the data cube is based on the sparse matrix structure concept: the missing values for each dimension are ignored, since they are empty values. Moreover, users do not performed the same number of actions and the completely empty slices of the cube may occur and ignored as well.

4   Experimental Evaluation

To validate the capability of our approach in discovering user profiles, we evaluate MDLDA on two datasets: the first is a real scenario while the second is synthetically generated. Real data are used to assess the capability of the framework in profiling users, exploiting simultaneously their observable-action set and environmental information. Synthetic data are exploited to test the anomaly detection capability of MDLDA.

4.1   Real Scenario

MovieLens1M[5] is the first dataset employed in our experimentation. Data are public and collected by "The GroupLens Research Project" developed within the Department of Computer Science and Engineering at the University of Minnesota [12]. This dataset contains 1,000,209 anonymous appreciation ratings on approximately 3,700 movies made by 6,040 users and it is widely used as a benchmark for several mining tasks.

MovieLens1M contains tuples $\langle u; i; r; t \rangle$, where u and i are, respectively, user and item identifiers, t is a timestamp, $r \in \{1, ..., 5\}$ is the rating value. The data collection period is between 1997 and 1998. To extract a useful sample for analyzing the model performances, data were filtered according to two criteria: *(i)* we removed user with less than 20 ratings and *(ii)* without demographical information.

Demographical information is *gender, age* and *occupation* while for each movie we know *title* and *genre*. In our case, these attributes represent the different dimensions that characterize user actions, in other words their preferences. The age values are discrete and are shown in Table 3.

Table 3. Discretization of the users' ages.

Age label	Real age
1	<18
18	18–24
25	25–34
35	35–44
45	45–49
50	50–55
56	>55

Occupation and genre are represented using codes. So, the fully-preprocessed dataset was obtained by combining user preferences, demographic information and information about movies. Following we report a table record as example:

---

[5] https://grouplens.org/datasets/movielens/1m/.

$u1; F; 45; 10; 2389; 0; 0; 0; 0; 0; 1; 0; 0; 0; 0; 1; 0; 0; 0; 0; 1; 0; 0; 4$

the first four attributes are the user's code, the gender, the age and the employment code, respectively. The value 2389 is the code of the movie (in this case Psycho 1998). The next sequence of eighteen numbers refers to the categories: if a category field is valued 1 the movie belongs to category, 0 otherwise. In this case, the categories valued with 1 are *Crime, Horror and Thriller*. The last value is the rating provided by this user: in this case four stars.

User Profiles Discovery Analysis. MovieLens1M have been used as case study to evaluate the proposed algorithm performance in discovering user profiles. In this application scenario, a behavioral profile identifies a group of users who share both a preference for a particular type of movie and additional features which can be exploited to predict their interests. It is worth to notice that, exploiting the profile, it is quite easy suggesting new movie to see, planning promotions, etc. The experiment consists of two steps. First, we apply the algorithm for the discovering behavioral profiles and a clustering algorithm for identifying the groups. Then, we analyzed the results by assigning semantics to the identified groups.

The algorithm MDLDA extracts a set of latent variables from the data, i.e. the links between the users and the provided preferences. Latent variables allow us to associate the users to the movies which can meet their interests. In particular, the aim of this analysis is to understand the features that the users, belonging to a particular group, share. MDLDA finds for each user a behavior profile presented by a probability distribution over the topics, so, in order to cluster the users, we used the Expectation-Maximization (EM) clustering algorithm [8] with a random initialization over the profiles. We identified 10 user's groups which exhibit similar behaviors and exploiting the result of the MDLDA algorithm we have been able to discriminate the representative features for each cluster.

In Table 4, we report the parameter setting for MDLDA algorithm according to the experience in [18].

For the clustering step, the EM algorithm was executed with 10 clusters.

Table 4. Parameter setting.

Parameter	Value
α_k, with $k \in \{1, \ldots, K\}$	2
$\beta_{k(d)}$ with $k \in \{1, \ldots, K\}$ and $d \in \{1, \ldots, n\}$	0.1
Max number of iterations	800
burnIn	300
sampleLag	25
Max number of α update iterations	30

Table 5. Latent variables - clusters

LV/Cluster	C1	C2	C3	C4	C5	C6	C7	C8	C9	C10
LV#1	0.0167	0.0155	0.016	0.0171	0.017	0.0162	**0.1291**	0.0142	0.0164	0.0164
LV#2	0.0175	0.0155	0.0161	0.017	0.017	0.016	0.0787	0.0143	0.0165	0.0164
LV#3	0.0169	0.0155	0.0167	**0.6425**	0.017	0.0165	0.0165	0.0142	0.0164	0.0164
LV#4	0.0169	0.0155	0.016	0.017	0.017	0.0161	0.0162	**0.6963**	0.0164	0.0164
LV#5	0.0168	0.0158	0.0166	0.0177	0.0179	0.016	**0.1061**	0.0142	0.0165	0.0168
LV#6	0.0183	0.0164	0.016	0.017	0.0171	**0.209**	0.0164	0.0144	0.0174	0.0165
LV#7	0.0169	0.0155	0.0169	0.017	0.017	0.0159	**0.0947**	0.0142	0.0164	0.0166
LV#8	0.0168	**0.6478**	0.0162	0.0172	0.017	0.0159	0.0162	0.0142	0.0164	0.0165
LV#9	0.0168	0.0158	0.016	0.017	0.017	0.0165	0.0163	0.0142	**0.6191**	0.0166
LV#10	0.0179	0.0157	**0.6258**	0.017	0.017	0.016	0.0164	0.0142	0.0168	0.0164
LV#11	0.057	0.0703	0.0826	0.0495	0.0218	0.0755	0.0545	0.0446	0.0833	0.0644
LV#12	0.0167	0.0157	0.016	0.0174	0.0174	0.0161	0.0162	0.0142	0.0164	0.6378
LV#13	**0.6357**	0.0159	0.016	0.017	0.017	0.016	0.0162	0.0159	0.0164	0.0169
LV#14	0.0168	0.0155	0.016	0.017	0.6701	0.016	0.0163	0.0148	0.0164	0.0164
LV#15	0.0168	0.0155	0.0168	0.017	0.017	0.0159	0.0893	0.0142	0.017	0.0164
LV#16	0.0167	0.0156	0.0167	0.017	0.017	**0.2355**	0.0163	0.0142	0.0164	0.0164
LV#17	0.0181	0.0156	0.016	0.017	0.017	0.0159	**0.0954**	0.0153	0.0164	0.0165
LV#18	0.0167	0.0155	0.016	0.017	0.0172	0.0159	0.0874	0.0142	0.0164	0.0164
LV#19	0.0167	0.0155	0.016	0.017	0.0171	0.016	0.0853	0.0142	0.0164	0.0173
LV#20	0.0172	0.0157	0.016	0.0179	0.0174	**0.223**	0.0163	0.0142	0.0165	0.0167

Evaluation Results. In this section we analyzed the clustering results. First, we selected the most influential latent variables for each cluster and then, for each latent variable we identified the user attributes and the predominant genres.

In Table 5, we show for each cluster the most characterizing latent variables. In some cases, we can note that a cluster is denoted by more latent variables.

In Table 6 we show the distribution of the instances for each cluster. A preliminary analysis allows to identify a larger cluster compared with the others, which exhibit similar sizes.

A more detailed analysis, based on the discriminative latent variables, permits to associate interesting knowledge to the clusters. The identified features don't allow to model all users, but are discriminative for the groups. For example, cluster 1 is composed by male users (59% of the cluster), in particular students aged between 18 and 24 years. The mainly genres watched by these users are drama and comedy. Another group of users (cluster 4) contains women aged between 25 and 34 years employed in the health sector. The preferred genre is romance. The cluster 2 is characterized by male users (68% of the cluster) between 25 and 34 years who prefer action movies. Many users belonging to this cluster declared to be artists. Another interesting group is cluster 9. In this case, the 80% of the users are young men (25–34), who work as programmers and watch thriller movies. The largest cluster is made up of retirees, teachers or engineers and most part of users belong mainly to two age groups: 25–34 years and over-56. It is the largest cluster so it is difficult to identify a predominant gender (male/female). In this case, comedy is the most watched genre.

These cohesive results prove the effectiveness of the algorithm in profiling and identifying groups of users that exhibit similar behavior. The obtained profiles describe mainly young, with high level of education and who are big "consumers" of movies. This demographic composition depends on how the data were collected.

Table 6. Cluster distributions

ClusterId	Distribution
C1	410 (7%)
C2	259 (4%)
C3	409 (7%)
C4	322 (5%)
C5	281 (5%)
C6	895 (15%)
C7	2443 (40%)
C8	414 (7%)
C9	319 (5%)
C10	288 (5%)

Finally, we tried to compare our solution with a standard clustering approach, specifically with the algorithm Expectation Maximization (EM). In order to make a fair comparison between the two techniques, we needed to preprocess our data pivoting the preferences in movielens on the user column, so to obtain a suitable and correct input for EM. The preprocessed table was sparse and exhibited a very high dimensionality (over 75 K attributes). Even if we employed several feature selection strategies, the EM algorithm (in different implementations) was unable to return a result in a reasonable time, against tens of minutes of our approaches. In the below table, we show how the proposed approach was able to provide the results varying the number of topics exhibiting a linear behavior (Table 7).

Table 7. Computation times of our algorithm on movielens dataset

Number of topics	Time (in min.)
5	12
10	23
15	35
20	44
50	118

4.2   Synthetic Scenario

In this section we describe the experiments on a synthetic dataset generated with different distribution and different level of noise. This set of experiments gives us the opportunity to understand how our approach is capable of detecting different types of anomalous behaviors, i.e. behaviors with multi-level differences, from actions quite close to the normal behavior to actions very different from normal ones.

The input data of our algorithm are human actions specified by a sequence of values, each one belonging to a specific context, i.e. the type of action (see Sect. 3). At the end of preprocessing process, all data are normalized and discretized, for this reason the synthetic dataset generation process produces records with nominal values with generic labels. The algorithm for the data generation can be divided in two phases. In the first phase, we generate the action templates that are the distinct records for all users and they are duplicated in the second phase to create the dataset. Each template describes an action of a user and it is defined by a tuple with a user ID, an action ID and all values of possible contexts. The number of contexts is a configuration parameter. In our experimentation we set this value to 20. Each template is equipped with a random non negative weight. Then, we sample (with replacement) actions from the template set according to the assigned weights. A template with high weight will be chosen more frequently than a low-weighted one. In the next phase, we add noise within the obtained actions in order to generate anomalies for each user.

Then, we assign a label to each action which represent the anomaly level. The label is the type of the action (i.e. normal vs anomaly) and it can have three possible values: 0, 1 and 2. With 0 we mark the action as normal. With 1 we mark the action as an anomalous one with x different contexts compared to a normal action (x is a configuration parameter) in the generated dataset. With 2 we mark the actions which strongly differ from the original template.

In the second phase, we resample the data previously generated according to three parameters: the number of normal actions, the number of anomaly actions of type 1 and the number of anomaly actions of type 2 to generate for each user.

This data generation is motivated by the fact that we can simulate users frequently doing typical actions and sometimes performing unexpected operations, like for instance a manager working on files in its office or a student studying at the school. The two types of anomaly action are required to model two different deviation from a normal behavior. With anomaly action of type 1, we model the case of an action that is considered almost normal with little variations, for instance a student that studying math during night time: most contexts are equal like in a normal action, but very few of them (the time) are different. The other type of anomaly is an action completely different from the normality.

Given a user u, we define the anomaly level of an action a according to a probabilistic framework:

$$\Pr(a|u) = \sum_{k=1}^{K} \Pr(k|u) \Pr(a|k)$$

$$= \sum_{k=1}^{K} \theta_{u,k} \prod_{d \in D} \Pr(i_{u,a}^{(d)}|k) \qquad (7)$$

$$= \sum_{k=1}^{K} \theta_{u,k} \prod_{d \in D} \phi_{k,i_{u,a}^{(d)}}.$$

The lower is the value of this probability the higher is the anomaly degree of the action for the target user:

$$degree(a, u) = 1 - \Pr(a|u). \qquad (8)$$

Table 8 contains the results of the anomaly detection experiments. We considered 1,000 users, for each user we generated 5 normal-action templates, 3 nearly normal action templates and 50 anomaly templates. Each row of the table represents a noise injection with the proportion 2:1 for (resp.) low and high anomaly templates. The first column represents the noise rate within the data, while the columns labelled "Gain $(y\%)$" represent the relative number of correctly predicted anomalies exploiting the first $y\%$ predictions of our model according to the anomaly degree (Eq. 8), in other words is a tabular representation of the Cumulative Gains Chart.

According to the obtained results, the model seems to show a quite good ability in detecting anomalies within the generated data.

Table 8. Cumulative gains with different noise level.

Noise %	Gain (5%)	Gain (10%)	Gain (15%)	Gain (20%)
5	42	48	57	61.8
10	45	49	54	59.9
25	24.4	37.5	41.7	45.9
35	15.7	26.6	29.7	33.3

5 Conclusion

In this paper we proposed a novel approach for the Human Behavior Computing based on Multidimensional Latent Factor Modeling. We defined an extension of the well-known approach LDA, namely Multidimensional Latent Dirichlet Allocation (MDLDA), capable of deal with arbitrary multivariate elements. An experimental experience has been shown in order to prove the utility of a multidimensional perspective on human behavior. This paper represents a preliminary work in this direction, that, with further investigation, seems to be a promising branch of research for a better fitting of models for human actions.

References

1. Agarwal, D., Chen, B.C.: Regression-based latent factor models. In: Proceedings of the 15th ACM SIGKDD International Conference on Knowledge Discovery and Data Mining, KDD 2009, NY, USA, pp. 19–28. ACM, New York (2009)
2. Barber, D.: Bayesian Reasoning and Machine Learning. Cambridge University Press, Cambridge (2012)
3. Barbieri, N., Costa, G., Manco, G., Ortale, R.: Modeling item selection and relevance for accurate recommendations: a Bayesian approach. In: Proceedings of the Fifth ACM Conference on Recommender Systems, RecSys 2011, NY, USA, pp. 21–28 (2011). http://doi.acm.org/10.1145/2043932.2043941
4. Barbieri, N., Manco, G., Ortale, R., Ritacco, E.: Balancing prediction and recommendation accuracy: hierarchical latent factors for preference data. In: Proceedings of the Twelfth SIAM International Conference on Data Mining, Anaheim, California, USA, 26–28 April 2012, pp. 1035–1046 (2012)
5. Blei, D.M., Ng, A.Y., Jordan, M.I.: Latent Dirichlet allocation. J. Mach. Learn. Res. **3**, 993–1022 (2003)
6. Cao, L., Yu, P.S.: Behavior Computing: Modeling, Analysis, Mining and Decision. Springer Publishing Company, Incorporated, Heidelberg (2014)
7. Cheng, J., Yuan, T., Wang, J., Lu, H.: Group latent factor model for recommendation with multiple user behaviors. In: Proceedings of the 37th International ACM SIGIR Conference on Research & Development in Information Retrieval, SIGIR 2014, NY, USA, pp. 995–998. ACM, New York (2014)
8. Dempster, A.P., Laird, N.M., Rubin, D.B.: Maximum likelihood from incomplete data via the EM algorithm. J. R. Stat. Soc. Ser. B **39**(1), 1–38 (1977)
9. Griffiths, T.L., Steyvers, M.: Finding scientific topics. Proc. Natl. Acad. Sci. **101**(suppl 1), 5228–5235 (2004)
10. Griffiths, T.: Gibbs sampling in the generative model of latent Dirichlet allocation (2002)
11. Guerzhoy, M., Hertzmann, A.: Learning latent factor models of travel data for travel prediction and analysis. In: Sokolova, M., Beek, P. (eds.) AI 2014. LNCS (LNAI), vol. 8436, pp. 131–142. Springer, Cham (2014). doi:10.1007/978-3-319-06483-3_12
12. Harper, F.M., Konstan, J.A.: The MovieLens datasets: history and context. ACM Trans. Interact. Intell. Syst. **5**(4), 191–1919 (2015)
13. Jamali, M., Ester, M.: A matrix factorization technique with trust propagation for recommendation in social networks. In: Proceedings of the Fourth ACM Conference on Recommender Systems, RecSys 2010, NY, USA, pp. 135–142. ACM, New York (2010)
14. Minka, T.: Estimating a Dirichlet distribution (2000)
15. Pritchard, J.K., Stephens, M., Donnelly, P.: Inference of population structure using multilocus genotype data. Genetics **155**(2), 945–959 (2000)
16. Skondral, A., Rabe-Hesketh, S.: Latent variable modelling: a survey. Scand. J. Stat. **34**(4), 712–745 (2007)
17. Urtasun, R., Darrell, T.: Discriminative Gaussian process latent variable model for classification. In: Proceedings of the 24th International Conference on Machine Learning, ICML 2007, NY, USA, pp. 927–934. ACM, New York (2007)
18. Hoffman, M., Blei, D.M., Bach, F.: Online learning for latent Dirichlet allocation. In: Advances in Neural Information Processing Systems, vol. 24 (2010)

Multi-view Approach to Parkinson's Disease Quality of Life Data Analysis

Anita Valmarska[1,2]([✉]), Dragana Miljkovic[1], Marko Robnik-Šikonja[3], and Nada Lavrač[1,2]

[1] Jožef Stefan Institute, Ljubljana, Slovenia
{anita.valmarska,dragana.miljkovic,nada.lavrac}@ijs.si
[2] Jožef Stefan International Postgraduate School, Ljubljana, Slovenia
[3] Faculty of Computer and Information Science, Ljubljana, Slovenia
marko.robnik@fri.uni-lj.si

Abstract. Parkinson's disease is a neurodegenerative disorder that affects people worldwide. While the motor symptoms such as tremor, rigidity, bradykinesia and postural instability are predominant, patients experience also non-motor symptoms, such as decline of cognitive abilities, behavioural problems, sleep disturbances, and other symptoms that greatly affect their quality of life. Careful management of patient's condition is crucial to ensure the patient's independence and the best possible quality of life. This is achieved by personalized medication treatment based on individual patient's symptoms and medical history. This paper explores the utility of machine learning to help development of decision models, aimed to support clinicians' decisions regarding patients' therapies. We propose a new multi-view methodology for determining groups of patients with similar symptoms and detecting patterns of medications changes that lead to the improvement or decline of patients' quality of life. We identify groups of patients ordered in accordance to their quality of life assessment and find examples of therapy modifications which induce positive or negative change of patients' symptoms. The results demonstrate that motor and autonomic symptoms are the most informative for evaluating the quality of life of Parkinson's disease patients.

Keywords: Multi-view learning · Parkinson's disease · Subgroup discovery · Rule learning · Personalized medicine

1 Introduction

Parkinson's disease is a neurodegenerative disorder that affects people worldwide. Due to the death of nigral neurons, there is a shortage of dopamine in human brain causing several motor symptoms: tremor, rigidity, bradykinesia and postural instability. In addition to motor symptoms, Parkinson's disease is associated also with non-motor symptoms, which include cognitive, behavioural, and autonomic problems. These symptoms significantly decrease the quality of life of both the patients affected by Parkinson's disease and their families.

© Springer International Publishing AG 2017
A. Appice et al. (Eds.): NFMCP 2016, LNAI 10312, pp. 163–178, 2017.
DOI: 10.1007/978-3-319-61461-8_11

Around 6.3 million people have the condition worldwide [1]. In Europe, more than one million people live with Parkinson's disease and this number is expected to double by 2030 [11]. Parkinson's disease is the second most common neurodegenerative disease (after Alzheimer's disease) and its prevalence continues to grow as the population ages. Currently, there is no cure for Parkinson's disease. The reasons for the cell death are still poorly understood. The management of symptoms is of crucial importance for patients' quality of life, mainly addressed with antiparkinson medication, such as levodopa and dopamine agonists.

While many different studies found in the literature address specific aspects of the disease, there are few research efforts that adopt a holistic approach to disease management [13]. The PERFORM [21], REMPARK [19] and SENSE-PARK [3] systems are intelligent closed-loop systems that seamlessly integrate a range of wearable sensors (mainly accelerometers and gyroscopes), constantly monitoring several motor signals of the patients and enabling the prescribing clinicians to remotely assess the status of the patients. Clinicians have a real time image of the patients' condition. Based on each patient's response to her therapy (manifested by the change of the motor symptoms), the prescribing physician is able to adjust medication schedules and personalize the treatment [13]. However, no data mining paradigms are used in the mentioned systems.

The PD_manager [2] EU Horizon 2020 project aims at developing an innovative, mobile-health, patient-centric platform for Parkinson's disease management. One of the PD_manager phases involves mining of data collected from Parkinson's disease patients in order to help to construct a decision support system assisting clinicians and patients in personalized disease management. Our goal is to develop a multi-view clustering methodology, which will—based on the patients' allocation to clusters at each time point and their history of medication therapies—be able to make suggestions about modifications of particular patient's therapy, with the aim to improve the patient's quality of life.

This paper present the idea of using multi-view clusters of short time series data as reference points through which a patient potentially moves as the diseases progresses. Each cluster groups patients-at-some-time-point with similar characteristics. The difference in feature values and medications of any given patients as she moves from one cluster to another represent possible causes for that move. Computing statistics of the moves allows us to infer significant features and relevant medication changes for group of patients and to suggest medication changes for individual patients.

After presenting the background and motivation, Sect. 3 describes the Parkinson's Progression Markers Initiative (PPMI) data [17], captured for monitoring the development of the Parkinson's disease, together with the medications used for symptoms control. Section 4 proposes the methodology for analyzing the Parkinson's disease data through multi-view clustering of short time series and connecting the changes of symptoms-based clustering of patients to the changes in medication therapies with the goal to find treatment recommendation patterns. Section 5 presents the results of data analysis, tested on two data set variants. Finally, Sect. 6 presents the conclusions and ideas for further work.

2   Background and Motivation

Multi-view learning is a machine learning technique whose aim is to build a model from multiple views (data sets) by considering the diversity of different views [28]. These views may be obtained from multiple sources or different feature subsets and describe the same set of examples. Co-training [6] is one of the earliest representatives of multi-view learning. This approach considers two views consisted of both labeled and unlabeled data. Using any labeled data, co-training constructs a separate classifier for each view. The most confident predictions of each classifier on the unlabeled data are then used to iteratively construct additional labeled training data.

Multi-view clustering is concerned with clustering of data by considering the information shared by each of the separate views. Many multi-view clustering algorithms initially transform the available views into one common subspace (early integration), where they perform the clustering process [28]. Chaudhuri et al. [8] propose method for multi-view clustering where the translation to a lower vector space is done by using Canonical Correlation Analysis (CCA). Tzortzis and Likas [22] propose a multi-view convex mixture model that locates clusters' representatives (exemplars) using all views simultaneously. These exemplars are identified by defining a convex mixture model distribution for each view. Cleuziou et al. [9] present a method where in each view they obtain a specific organization using fuzzy k-means [5] and introduce a penalty term in order to reduce the disagreement between organizations in the different views. Cai et al. [7] propose a multi-view k-means clustering algorithm for big data. The algorithm utilizes a common cluster indicator in order to establish common patterns across views.

Kumar and Daumé [16] apply the co-training principle [6] in unsupervised learning. Clustering is performed on both views. Afterwards, cluster points from one view are used to modify the clustering structure of the other view. Appice and Malerba [4] employ the co-training principle in the multi-view setting for process mining clustering. The above mentioned approaches presume that each of the respective views is capable of producing clusters of similar quality when considered separately. He et al. [15] do not make that presumption. They combine multiple views under a principled framework, CoNMF (Co-regularized Non-negative Matrix Factorization), which extends NMF (Non-negative matrix factorization) for multi-view clustering by jointly factorizing the multiple matrices through co-regularization. The matrix factorization process is constrained by maximizing the correlation between pairs of views, thus utilizing information from each of the considered views. CoNMF is a multi-view clustering approach with intermediate integration of views, where different views are fused during the clustering process. The co-regularization of each pair of views makes the clustering process more robust to noisy views. The decision to use the CoNMF approach in our work was made based on the presumptions of the algorithm and the availability of their Python code.

Symptoms of patients suffering from Parkinson's disease can be divided into several views. When these views are combined they offer a better image of the patients' condition. We believe that the usage of multi-view clustering on the Parkinson's disease data will be able to identify clusters of patients that share similar symptoms. All patients' symptoms are susceptible to change through time: the symptoms will change depending on the received therapies, development of the disease, every day habits, etc. This will eventually lead to patients' allocation in different clusters in different time points depending on the progression of the disease. By identifying the migration of patients from one cluster to another, modifications of the medication treatments will be suggested in order to keep the patients in the clusters where patients share symptoms that show good quality of life. In the following sections we present a brief description of the available Parkinson's disease data, the proposed methodology and data analysis results.

3   Data

In this study we use the PPMI[1] data collection [17] gathered in the observational clinical study to verify progression markers in Parkinsons disease. The PPMI data collection consists of data sets describing different aspects of the patients' daily life. Below we describe the selection of PPMI data used in the experiments.

3.1   PPMI Symptoms Data Sets

The medical condition and the quality of life of a patient suffering from Parkinson's disease can be determined using the Movement Disorder Society (MDS)-sponsored revision of the Unified Parkinson's Disease Rating Scale (MDS-UPDRS) [14]. It is a questionnaire consisting of 65 questions concerning the development of the disease symptoms. The MDS-UPDRS is divided into four parts. Part I consists of questions about the 'non-motor experiences of daily living'. These questions address complex behaviors, such as hallucinations, depression, apathy, etc., and patient's experiences of daily living, such as, sleeping problems, daytime sleepiness, urinary problems, etc. Part II expresses 'motor experiences of daily living'. This part of the questionnaire examines whether the patient experiences speech problems, the need for an assistance with the daily routines, such as eating or dressing, etc. Part III is retained as the 'motor examination', while Part IV concerns 'motor complications', which are mostly developed when the main antiparkinson drug levodopa is used for a longer time period. Each question is anchored with five responses that are linked to commonly accepted clinical terms: 0 = normal (patient's condition is normal, symptom is not present), 1 = slight (symptom is present and has a slight influence on

---

[1] Data used in the preparation of this article were obtained from the Parkinsons Progression Markers Initiative (PPMI) database (www.ppmi-info.org/data). For up-to-date information on the study, visit www.ppmi-info.org. PPMI—a public-private partnership—is funded by the Michael J. Fox Foundation for Parkinson's Research and funding partners. List of funding partners can be found at www.ppmi-info.org/fundingpartners.

the patient's quality of life), 2 = mild, 3 = moderate, and 4 = severe (symptom is present and severely affects the normal and independent functioning of the patient, i.e. her quality of life is significantly decreased).

The Montreal Cognitive Assessment (MoCA) [10] is a rapid screening instrument for mild cognitive dysfunction. It is a 30 point questionnaire consisting of 11 questions, designed to assess different cognitive domains: attention and concentration, executive functions, memory, language, visuoconstructional skills, conceptual thinking, calculations, and orientation.

The Scales for Outcomes in Parkinson's disease Autonomic (SCOPA-AUT) is a specific scale to assess autonomic dysfunction in Parkinson's disease patients [26]. The Physical Activity Scale for the Elderly (PASE) [27] is a questionnaire which is a practical and widely used approach for physical activity assessment in epidemiologic investigations. Cognitive categorization (COGCAT) is a questionnaire filled in by clinicians evaluating the cognitive state and possible cognitive decline of patients.

The above data sets are periodically updated, thus allowing clinicians to monitor patients' disease development through time. Answers to the questions from each questionnaire form the vectors of attribute values. All of the considered data sets consist of attributes with ordered values: larger values indicate worsening of the symptoms described by the attributes, denoting a decreased patient's quality of life, while the opposite is true for attributes from the MoCA and PASE data sets.

When considering the possibility of using a multi-view framework, the independence of the separate views should be inspected. In their work of 2008, Goetz et al. [14] stated that the four parts of the MDS-UPDRS scale are considered to be independent due to the fact that obtained reliable factor structures for each part with the comparative fit index > 0.90 for each part, which supports the use of sum scores for each part in preference to a total score of all parts.

3.2 PPMI Concomitant Medications Log

The PPMI data collection offers information about all of the concomitant medications patients used during their involvement in the study. These medications are described by their name, the medical condition they are prescribed for, and when the patient has started and (if) ended the therapy. For the purpose of our research, we initially concentrate on whether the patient receives a therapy with antiparkinson medications, and which combination of antiparkinson medications she has received between each of the time points when the MDS-UPDRS test and the MoCA test have been administered. The main families of drugs used for treating motor symptoms are levodopa, dopamine agonists and MAO-B inhibitors [12]. Medications which treat Parkinson's disease related symptoms but are not from the above mentioned groups of medications are referred to as *other*.

3.3   Experimental Data Sets

Symptoms of patients suffering from Parkinson's disease are grouped into several data sets, representing distinct views of the data. These views consist of data from MoCA test, motor experiences of daily living, non-motor experiences of daily living, complex motor examination data, etc. For each patient these data are obtained and updated periodically (on each patient's visit to the clinician's office)—in the beginning of the patient's involvement in the PPMI study, and approximately every 6 months, in total duration of 5 years—providing the clinicians with the opportunity to follow the development of the disease. The visits of patient can be viewed as time points, and the collected data on each visit is data about the patient in the respective time point. All time points collected for one patient form a short time series. The experiments will address two settings: the analysis of *merged symptoms data* and the analysis of *multi-view symptoms data*.

The *merged symptoms data* are represented in a single data table, constructed by using the sums of values of attributes of the following data sets: MDS-UPDRS Part I (subpart 1 and subpart 2), Part II, Part III, MoCA, PASE, SCOPA-AUT, and COGSUM. Table 1 outlines the attributes used to construct the merged symptoms data, together with their range of values. This is a simplified representation using eight attributes, each representing the severity of symptoms of a given symptoms group, which proved to be valuable in initial experiments [24].

The *multi-view data* consist of eight data sets: MDS-UPDRS Part I, Part Ip, Part II, Part III, MoCA, SCOPA-AUT, PASE, and COGCAT. Each of these data sets consists of values of attributes, which represent answers to the questions from a particular questionnaire. In each data set, we included an additional attribute, which is the sum of values of attributes of the given data set (this equals the values of individual attributes used in the merged symptoms data).

Table 1. List of attributes used in the merged symptoms data set.

Dataset	Attribute name	Value range	Increased value decreases quality of life?
MDS-UPDRS Part I	NP1SUM	0–24	Yes
MDS-UPDRS Part Ip	NP1PSUM	0–28	Yes
MDS-UPDRS Part II	NP2SUM	0–52	Yes
MDS-UPDRS Part III	NP3SUM	0–138	Yes
MoCA	MCATOT	0–30	No
PASE	PASESUM	0–24	No
SCOPA-AUT	SCAUSUM	0–63	Yes
COGCAT	COGSUM	2–9	Yes

4   Methodology

To assist the clinicians in making decisions regarding the patients' therapy, we propose a method which involves a combination of multi-view clustering on patients' symptoms data and the analysis of histories of patients' medication treatments. Figure 1 shows an outline of the proposed methodology, which addresses changes of data over time (i.e. over several patient's visits) with the goal to suggest possible modifications of the medication treatment. The input to the methodology are PPMI data sets of patient symptoms (described in Sect. 3.1) and the PPMI medications log data set (described in Sect. 3.2), and the output are treatment recommendation patterns, which can be used to assist the clinician in deciding about further treatment of a patient.

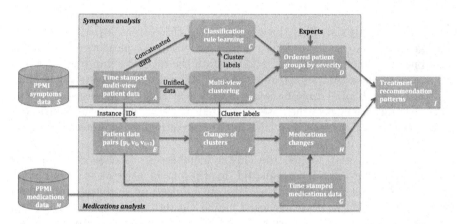

Fig. 1. Outline of the multi-view approach to Parkinson's disease quality of life data analysis.

The methodology consists of two separate threads whose outputs are combined to identify treatment recommendation patterns. The first thread, referred to as *Symptoms Analysis* in the top part of Fig. 1, concerns finding groups of patients with similar symptoms by grouping of patient instances, defined as (patient, visit) pairs, through multi-view clustering and describing these groups with induced classification rules from cluster labeled data instances. The second thread, referred to as *Medication analysis* in the bottom part of Fig. 1, concerns finding changes of medications and their dosages based on patients' symptoms changes between two consecutive visits to the clinician's (e.g., disease aggravation, improvement or no change). In this thread we observe patients moving from one cluster to another cluster in two consecutive time points, i.e. in two consecutive (patient, visit) pairs. The outcomes of the two threads are combined to a set of treatment recommendation patterns (i.e. increased/decreased/unchanged dosage of medications, for the four groups of medications, mentioned in Sect. 3.2).

Initially, we construct patient-visit pairs (p_i, v_{ij}) based on the patients and their visits to the clinician. For each patient p_i a set of pairs (p_i, v_{ij}) is constructed, where v_{ij} corresponds to an individual visit, and v_{ij} and v_{ij+1} correspond to two consecutive patient's visits. This is followed by clustering of the available data. These patient-visit pairs, called *instances*, represent the basic unit of analysis in the *Symptoms Analysis* thread of the methodology. The instance's (p_i, v_{ij}) attribute values correspond to symptoms severity of patient p_i on visit j. On the other hand, the basic unit of analysis in the *Medications Analysis* thread of the methodology are $(p_i, v_{ij}, c_{ij}, m_{ij}, v_{ij+1}, c_{ij+1}, m_{ij+1})$ tuples, where c_{ij} is the cluster label for instance (p_i, v_{ij}) and m_{ij} are the medications the patient p_i takes at the time of visit j. Elements c_{ij+1} and m_{ij+1} are the cluster assignment and medications information about the same patient on visit $j + 1$, i.e. at the time of the next visit.

Symptoms Analysis. This thread consists of three steps. First, we perform clustering on instances in order to determine groups of patients with similar symptoms. Our methodology can address both the merged symptoms data and the multi-view data analysis setting. The only difference is the clustering method applied in step B of the methodology outlined in Fig. 1. For clustering of the multi-view data, which is the case illustrated in Fig. 1, we use the multi-view clustering approach proposed in [15]. In the case of merged symptoms data, we perform k-means clustering.

In the second step, we use the cluster membership as class labels in rule learning in order to obtain meaningful descriptions of patients in each cluster (step C). Rule sets describing the data are induced on a concatenated data set consisting of data sets considered in the clustering step B. In this work, the rule sets for each class variable are learned using our recently developed DoubleBeam-RL algorithm [23, 25]. This is a separate-and-conquer rule learning algorithm which uses two beams and separate heuristics for rule refinement and rule selection. By using heuristics that take full advantage of the refinement and selection process separately, the algorithm is able to find rules which maximize the number of covered positive examples and minimize the number of covered negative examples - which is the goal of classification rule learning algorithms [20]. In the third step, expert knowledge was used to interpret the obtained groups of patients and to order them according to the severity of symptoms exhibited by patient instances assigned to them.

Based on the experts' interpretation of clusters, we consider cluster changes to be either positive or negative. When patient moved from a cluster described by symptoms indicating worse quality of life to one depicted by better quality of life indicators, we consider this change to be positive. A negative cluster change occurs when the symptoms of a patient worsened. In step I we combine detected medications changes from step H and cluster severity information from step D. The combined information contains medications changes for positive cluster changes and for negative cluster changes i.e. medication changes with improvement or aggravation of the patients' symptoms.

Medications Analysis. In this thread of the methodology we determine the medications changes that have occurred simultaneously to moves between clusters observed in patients during two consecutive time points (two consecutive visits). An important benefit of our approach is that each patient provides a context for herself. By following the development of symptoms for each patient separately, we remove the influence of any chronic conditions the patient is treated for.

Information about patients' assignment to clusters and their medication therapy in two consecutive visits is held in $(p_i, v_{ij}, c_{ij}, m_{ij}, v_{ij+1}, c_{ij+1}, m_{ij+1})$ tuples. In step $\boldsymbol{H}$ we follow all patients through time. For each pair of patients' consecutive visits to the clinician, we record the cluster change that has occurred between the two visits, $c_{ij} \rightarrow c_{ij+1}$, as well as the change in medications prescriptions, $m_{ij} \rightarrow m_{ij+1}$, which patients received in the consecutive time points. For each antiparkinson drug group (levodopa, dopamine agonists, MAOB inhibitors, and others) we record whether their dosage has increased, decreased or stayed unchanged between the two visits. Dosages of PD medications are translated into a common Levodopa Equivalent Daily Dosage (LEDD) which allows for comparison of different therapies (different medications with personalized daily plans of intake). Based on the clusters ordered by the experts and the information of medications change for each cluster change, we determine patterns of medications adaptations, which have resulted in patients' symptoms' improvement or aggravation.

5   Data Analysis

The experiments are divided into two parts: the analysis of merged symptoms data and the analysis of multi-view symptoms data, briefly described in Sect. 3.3. As described in Sect. 4, the merged symptoms data are clustered using k-means, where we set the number of clusters to 3 by using the silhouette analysis technique [18] and by manual inspection of the silhouette graphs, while clustering of multi-view data is performed using the CoNMF approach [15], explained in Sect. 2.

5.1   Results of Merged Symptoms Data Analysis

To determine the patients' symptoms evolution, for each Parkinson's disease patient considered in our data we investigate how the clusters in which the patient was involved have changed between two consecutive time points. When a patient has moved from cluster with a lower index to one with higher index, we note that the patient's symptoms have worsen and thus consider this change to be negative. A positive cluster change is recorded if patient's symptoms have improved and the patient has moved to a cluster with smaller index. The medications change patterns for positive and negative cluster change are obtained by using the methods described in Sect. 4.

Table 2 outlines rules describing the clusters obtained from the merged symptoms data analysis. These rules indicate that the clusters are ordered and contain instances (patient at certain time point) with different severity of their motor symptoms. *Cluster 0* consists of instances with sum of motor symptoms severity up to 22 (out of 138). Patients that have slightly worse motor symptoms are assigned to *cluster 1* (sum of motor symptoms severity between 23 and 42). In *cluster 2* there are patients whose motor symptoms significantly affect their motor functions (sum of motor symptoms severity greater than 42). The worsening of motor symptoms is followed by aggravation of non-motor symptoms, mostly autonomic symptoms (sleeping, urinary, or constipation problems). This can be observed by the increased values of attributes SCAUSUM and NP2SUM in rule sets describing *cluster1* and *cluster2*.

Table 2. Rules describing clusters obtained by k-means clustering on the unified data set of attribute sums. Variables p and n denote the number of covered true positive and false positive examples respectively.

Rule		p	n
Rules for cluster 0			
NP3SUM $\leq$ 20	$\leftarrow$ cluster = 0	488	4
NP3SUM $\leq$ 21 AND NP2SUM $\leq$ 6	$\leftarrow$ cluster = 0	321	0
NP3SUM = (19, 22] AND NP1SUM = 0	$\leftarrow$ cluster = 0	54	23
Rules for cluster 1			
NP3SUM = (22,30]	$\leftarrow$ cluster = 1	323	13
NP3SUM = (30, 39] AND SCAUSUM = (4, 10]	$\leftarrow$ cluster = 1	91	17
NP3SUM = (22, 42] AND NP2SUM = (0, 6]	$\leftarrow$ cluster = 1	206	6
NP3SUM = (22, 34] AND SCAUSUM = (10, 17] AND PASESUM >9	$\leftarrow$ cluster = 1	101	6
Rules for cluster 2			
NP3SUM >42	$\leftarrow$ cluster = 2	125	1
NP3SUM >37 AND NP1PSUM >5 AND MCAVFNUM $\leq$ 18	$\leftarrow$ cluster = 2	123	6
NP3SUM >30 AND NP2SUM >17	$\leftarrow$ cluster = 2	82	0
SCAUSUM >20 AND NP2SUM >9 AND MCAVFNUM $\leq$ 24	$\leftarrow$ cluster = 2	54	18
NP3SUM >30 AND SCAUSUM >11 AND NP2SUM >12	$\leftarrow$ cluster = 2	123	2
NP3SUM >36 AND SCAUSUM >6 AND NP2SUM >6 AND NP1PSUM >2	$\leftarrow$ cluster = 2	168	6

Figure 2 indicates that patients' motor symptoms improve when the dosage of medications from the levodopa drug group is increased and the dosage of dopamine agonists is decreased or stays the same. When the dosage of both levodopa medications and dopamine agonists is increased the motor symptoms of the patients worsen. Clinicians prescribe and gradually increase dosages of levodopa to handle the motor symptoms of patients. The usage of inadequately high dosages of dopamine agonists produces side effects affecting the non-motor symptoms of patients. A decrease of dosage eliminates these side effects and improves the patient's status. Figure 2(a) presents the medications changes when a positive cluster change has occurred. Red bars represent the number of times the dosage of medications from certain medication group has increased. Similarly, the number of times the medication dosage has decreased is shown with green. Blue bars present the number of times when a positive cluster change has occurred, but the medication dosage has stayed unchanged. Figure 2(b) outlines the medications changes when a negative cluster change has taken place.

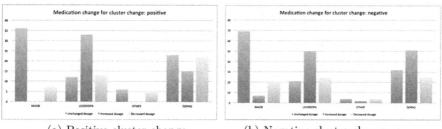

(a) Positive cluster change.   (b) Negative cluster change.

Fig. 2. Recorded Parkinson's disease medication change when patient's cluster allocation has changed. Clusters were obtained from merged symptoms data set. Positive cluster change indicates that patient's symptoms improved. A negative cluster change occurs when patient's symptoms worsen. (Color figure online)

5.2   Multi-view Results

Table 3 outlines rules describing the first cluster obtained from the applied multi-view clustering method. Attribute names (their IDs in the PPMI data collection) are written in parenthesis. Rules presented in Table 3 contain attributes from the following data sets: MDS-UPDRS Part Ip (NP1LTHD), MDS-UPDRS Part II (NP2TRMR, NP2HYGN, NP2TURN), MDS-UPDRS Part III (DYSKPRES, NP3FTAPR), MOCA (MCADATE, MCASER7, MCABDS, MCAVF, MCA-CLCKH, MCACITY, MCALION, MCAABSTR, MCAABDS), SCOPA-AUT (SCAUSUM, SCAU3, SCAU4, SCAU7, SCAU8, SCAU9, SCAU10, SCAU11, SCAU13, SCAU15), PASE (LAWNWRK), and COGCAT (FNCDTCOG).

Similar to the results from the merged view clustering approach, clusters obtained by the multi-view approach can be ordered by the severity of symptoms of patients assigned to them. In *cluster 0* there are patients whose symptoms are most tolerable, while in *cluster 2* are patients experiencing symptoms that significantly affect their motor functions. *Cluster 1* contains patients with symptoms worse than patients assigned to *cluster 0*, but better than symptoms of patients involved in *cluster 2*. A post analysis has revealed that there are no significant intersections of instances assigned to clusters in the merged view approach and instances assigned to clusters in the multi-view approach.

Clusters obtained by the multi-view approach are described with specific rules. In addition to the sums of symptoms severity, we were able to determine

Table 3. Rules describing *cluster*1 obtained by the multi-view clustering approach proposed in [15]. Variables p and n denote the number of covered true positive and false positive examples respectively.

Rule		p	n
IF: SCAUSUM $\leq$ 0	$\leftarrow$ cluster = 1	18	0
ELSE IF: SCAUSUM $\leq$ 4 AND (SCAU13) PASS URINE AT NIGHT = 2	$\leftarrow$ cluster = 1	26	0
ELSE IF: (SCAU8) DIFFICULTY RETAINING URINE = 0 AND SCAUSUM $\leq$ 3 AND (SCAU10) AFTER PASSING URINE BLADDER NOT COMPLETELY EMPTY = 0 AND (SCAU13) PASS URINE AT NIGHT = 1 AND (NP1LTHD) LIGHTHEADEDNESS ON STANDING = 0	$\leftarrow$ cluster = 1	41	0
ELSE IF: (FNCDTCOG) FUNCTIONAL IMPAIRMENT DUE TO COGNITIVE = 0 AND (MCADATE) ORIENTATION - DATE = 1 AND (MCASER7) ATTENTION - SERIAL 7S = 3 AND SCAUSUM = (8,15] AND (SCAU13) PASS URINE AT NIGHT = 3 AND (MCABDS) ATTENTION - BACKWARD DIGIT SPAN = 1 AND (SCAU15) LIGHT-HEADED FOR SOME TIME AFTER STANDING = 0	$\leftarrow$ cluster = 1	50	6
ELSE IF: (MCAVF) VERBAL FLUENCY = 1 AND SCAUSUM $\leq$ 7 AND (SCAU13) PASS URINE AT NIGHT = 2 AND (NP2TRMR) REST TREMOR AMPLITUDE - LLE = 0	$\leftarrow$ cluster = 1	46	5
ELSE IF: (MCACLCKH) VISUOCONSTRUCTIONAL SKILLS (CLOCK HANDS) = 1 AND (MCADATE) ORIENTATION - DATE = 1 AND (MCAVF) VERBAL FLUENCYNUM $\leq$ 26 AND (DYSKPRES) WERE DYSKINESIAS PRESENT = 0 AND (MCACITY) ORIENTATION - CITY = 1 AND (MCALION) NAMING - LION = 1 AND (SCAU9) INVOLUNTARY LOSS OF URINE = 0 AND (SCAU7) INVOLUNTARY LOSS OF STOOLS = 0 AND (SCAU3) FOOD STUCK IN THROAT = 0 AND SCAUSUM = (7,10] AND (SCAU13) PASS URINE AT NIGHT = 2 AND (MCAABSTR) ABSTRACTION = 2 AND (MCABDS) ATTENTION - BACKWARD DIGIT SPAN = 1	$\leftarrow$ cluster = 1	28	10
ELSE IF: (SCAU4) FULL VERY QUICKLY DURING A MEAL = 0 AND SCAUSUM $\leq$ 5 AND (SCAU11) WEAK STREAM OF URINE = 0 AND (SCAU13) PASS URINE AT NIGHT = 1 AND (NP3FTAPR) FINGER TAPPING RIGHT HAND = 1 AND (LAWNWRK) LAWN WORK = 2	$\leftarrow$ cluster = 1	13	3
ELSE IF: SCAUSUM = (8,15] AND (SCAU13) PASS URINE AT NIGHT = 3 AND (NP2HYGN) HYGIENE = 0 AND (NP2TURN) TURNING IN BED = 1	$\leftarrow$ cluster = 1	8	3

specific symptoms which describe patients' status. Due to space restrictions we only outline rules describing *cluster 1*.

Figure 3(a) presents the medications changes when the symptoms of a patient improved and a positive cluster change occurred. Figure 3(b) outlines the medications changes when a negative cluster change took place.

Results from the medications change pattern analysis in the multi-view clustering setting are similar to those obtained in the merged view setting—the improvement of patients' symptoms takes place when the dosage of medications

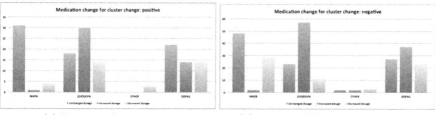

(a) Positive cluster change.   (b) Negative cluster change.

Fig. 3. Recorded Parkinson's disease medication change when patient's cluster allocation has changed. Clusters were obtained with multi-view clustering. Positive cluster
change indicates that patient's symptoms have improved. A negative cluster change
occurs when patient's symptoms worsen.

from the levodopa medications group increased and the dosage of dopamine agonists decreased or stayed the same. When patients move to a cluster with worse
symptoms, the change in the treatment plan is characterized by the increased
dosage of both levodopa medications and dopamine agonists. Increased dosages
of dopamine agonists can cause worsening of patients' cognitive symptoms.
Figures 2 and 3 outline that the dosage of MAOB inhibitors does not have any
significant influence on the change of the symptoms. In all observed cases when a
cluster change occurred, dosage of MAOB inhibitors remained the same. This is
in accordance with the Parkinson's disease literature and the clinicians' practice
of medication prescription for Parkinson's disease treatment.

6   Conclusions

The aim of our research is to develop a methodology which will make suggestions to clinicians about possible treatment changes that will improve the
patient's quality of life. We present the results from the proposed methodology
obtained by rule learning and clustering on patients' data. The results confirm
known facts about the Parkinson's disease: the motor symptoms, tremor, shak-
ing, involuntary movement, etc. are the characteristic symptoms of the disease
and significantly affect the quality of life of the suffering patient. Our experimental work also revealed that the autonomic symptoms (SCOPA-AUT) are very
informative when dividing patients into groups defined by patients' quality of
life evaluation. We show that Parkinson's disease patients can be divided into
clusters ordered in accordance to the severity of their symptoms. By following
the evolution of symptoms for each patient separately, we were able to determine
patterns of medications change which can lead to the improvement or worsening
of the patients' quality of life.

The rules describing the obtained clusters were either too general (merged
view setting) or too specific (multi-view setting) and may not be of sufficient
assistance to the clinicians. This is due to the nature of the used data, i.e. a vector of attribute sums (merged view) or a broad vector of attributes with numeric

values. In future work we will test our methodology with only a handful of carefully chosen attributes. These attributes, selected with the help of Parkinson's disease specialists, will be described by nominal values used in their everyday practice. We believe that by manually decreasing the attribute space we will be able to obtain description of groups of patients which are meaningful and helpful for the clinicians. Additionally, we will improve the medications suggestion process to produce numerical suggestions of drugs' dosage which should be prescribed to patients. Furthermore, we plan to formalize the patients' cluster changes and define a probabilistic model. The same approach will be applied for the determination of medications changes, which will allow more personalized approach for change of treatment recommendation. We will also explore the possibility of comparing the state of the proposed framework in a given time point with all of its past time points.

Acknowledgements. This work was supported by the PD_manager project, funded within the EU Framework Programme for Research and Innovation Horizon 2020 grant 643706. We acknowledge the support of the Slovenian Research Agency and the European Commission through The Human Brain Project (HBP), grant FP7-ICT-604102.

References

1. European Parkinson's Disease Association. http://www.epda.eu.com/. Accessed 01 July 2016
2. PD_manager: m-Health platform for Parkinson's disease management. EU Framework Programme for Research and Innovation Horizon 2020, Grant number 643706, 2015–2017. http://www.parkinson-manager.eu/
3. SENSE-PARK. Project's website. http://www.sense-park.eu/. Accessed 01 July 2016
4. Appice, A., Malerba, D.: A co-training strategy for multiple view clustering in process mining. IEEE Trans. Serv. Comput. **9**(6), 832–845 (2016)
5. Bezdek, J.C.: Pattern Recognition with Fuzzy Objective Function Algorithms. Plenum Press, New York (1981)
6. Blum, A., Mitchell, T.: Combining labeled and unlabeled data with co-training. In: Proceedings of the Eleventh Annual Conference on Computational Learning Theory. COLT 1998, pp. 92–100. ACM, New York (1998)
7. Cai, X., Nie, F., Huang, H.: Multi-view k-means clustering on big data. In: IJCAI 2013, Proceedings of the 23rd International Joint Conference on Artificial Intelligence, Beijing, China, August 3–9, 2013, pp. 2598–2604 (2013)
8. Chaudhuri, K., Kakade, S.M., Livescu, K., Sridharan, K.: Multi-view clustering via canonical correlation analysis. In: Proceedings of the 26th Annual International Conference on Machine Learning, ICML 2009, pp. 129–136 (2009)
9. Cleuziou, G., Exbrayat, M., Martin, L., Sublemontier, J.: Cofkm: a centralized method for multiple-view clustering. In: The Ninth IEEE International Conference on Data Mining, ICDM 2009, Miami, Florida, USA, 6–9, pp. 752–757 (2009)
10. Dalrymple-Alford, J., MacAskill, M., Nakas, C., Livingston, L., Graham, C., Crucian, G., Melzer, T., Kirwan, J., Keenan, R., Wells, S., et al.: The MoCA: well-suited screen for cognitive impairment in Parkinson disease. Neurology **75**(19), 1717–1725 (2010)

11. Dorsey, E., Constantinescu, R., Thompson, J., Biglan, K., Holloway, R., Kieburtz, K., Marshall, F., Ravina, B., Schifitto, G., Siderowf, A., et al.: Projected number of people with Parkinson disease in the most populous nations, 2005 through 2030. Neurology **68**(5), 384–386 (2007)
12. N.C.C. for Chronic Conditions (UK et al. Symptomatic pharmacological therapy in Parkinsons disease) (2006)
13. Gatsios, D., Rigas, G., Miljkovic, D., Seljak, B.K., Bohanec, M.: m-health platform for Parkinson's disease management. In: Proceedings of 18th International Conference on Biomedicine and Health Informatics CBHI (2016)
14. Goetz, C.G., Tilley, B.C., Shaftman, S.R., Stebbins, G.T., Fahn, S., Martinez-Martin, P., Poewe, W., Sampaio, C., Stern, M.B., Dodel, R., et al.: Movement disorder society-sponsored revision of the unified Parkinson's disease rating scale (MDS-UPDRS): scale presentation and clinimetric testing results. Mov. Disord. **23**(15), 2129–2170 (2008)
15. He, X., Kan, M.-Y., Xie, P., Chen, X.: Comment-based multi-view clustering of web 2.0 items. In: Proceedings of the 23rd International Conference on World Wide Web, pp. 771–782. ACM (2014)
16. Kumar, A., Daumé, H.: A co-training approach for multi-view spectral clustering. In: Proceedings of the 28th International Conference on Machine Learning, ICML, pp. 393–400 (2011)
17. Marek, K., Jennings, D., Lasch, S., Siderowf, A., Tanner, C., Simuni, T., Coffey, C., Kieburtz, K., Flagg, E., Chowdhury, S., et al.: The Parkinson's progression markers initiative (PPMI). Progress Neurobiol. **95**(4), 629–635 (2011)
18. Rousseeuw, P.J.: Silhouettes: a graphical aid to the interpretation and validation of cluster analysis. J. Comput. Appl. Math. **20**, 53–65 (1987)
19. Samà, A., Pérez-López, C., Rodríguez-Martín, D., Moreno-Aróstegui, J.M., Rovira, J., Ahlrichs, C., Castro, R., Cevada, J., Graça, R., Guimarães, V., et al.: A double closed loop to enhance the quality of life of Parkinson's disease patients: REM-PARK system. Innov. Med. Healthcare **207**, 115 (2015)
20. Stecher, J., Janssen, F., Fürnkranz, J.: Separating rule refinement and rule selection heuristics in inductive rule learning. In: Calders, T., Esposito, F., Hüllermeier, E., Meo, R. (eds.) ECML PKDD 2014. LNCS, vol. 8726, pp. 114–129. Springer, Heidelberg (2014). doi:10.1007/978-3-662-44845-8_8
21. Tzallas, A.T., Tsipouras, M.G., Rigas, G., Tsalikakis, D.G., Karvounis, E.C., Chondrogiorgi, M., Psomadellis, F., Cancela, J., Pastorino, M., Waldmeyer, M.T.A., et al.: PERFORM: a system for monitoring, assessment and management of patients with Parkinson's disease. Sensors **14**(11), 21329–21357 (2014)
22. Tzortzis, G., Likas, A.: Convex mixture models for multi-view clustering. In: Alippi, C., Polycarpou, M., Panayiotou, C., Ellinas, G. (eds.) ICANN 2009. LNCS, vol. 5769, pp. 205–214. Springer, Heidelberg (2009). doi:10.1007/978-3-642-04277-5_21
23. Valmarska, A., Lavrač, N., Fürnkranz, J., Robnik-Šikonja, M.: Refinement and selection heuristics in subgroup discovery and classification rule learning (2017). (under review)
24. Valmarska, A., Miljkovic, D., Lavrač, N., Robnik-Šikonja, M.: Towards multi-view approach to Parkinson's disease quality of life data analysis. In: 5th International Workshop on New Frontiers in Mining Complex Patterns at ECML-PKDD2016 (2016)
25. Valmarska, A., Robnik-Šikonja, M., Lavrač, N.: Inverted heuristics in subgroup discovery. In: Proceedings of the 18th International Multiconference Information Society (2015)

26. Visser, M., Marinus, J., Stiggelbout, A.M., Van Hilten, J.J.: Assessment of autonomic dysfunction in Parkinson's disease: the SCOPA-AUT. Mov. Disord. **19**(11), 1306–1312 (2004)
27. Washburn, R.A., Smith, K.W., Jette, A.M., Janney, C.A.: The physical activity scale for the elderly (PASE): development and evaluation. J. Clin. Epidemiol. **46**(2), 153–162 (1993)
28. Xu, C., Tao, D., Xu, C.: A survey on multi-view learning. arXiv preprint arXiv:1304.5634 (2013)

Pattern Discovery

Subgraph Mining for Anomalous Pattern Discovery in Event Logs

Laura Genga[1]([⊠]), Domenico Potena[1], Orazio Martino[1], Mahdi Alizadeh[2], Claudia Diamantini[1], and Nicola Zannone[2]

[1] Dipartimento di Ingegneria dell'Informazione, Università Politecnica delle Marche,
Ancona, Italy
{l.genga,d.potena,o.martino,c.diamantini}@univpm.it
[2] Eindhoven University of Technology, Eindhoven, The Netherlands
{m.alizadeh,n.zannone}@tue.nl

Abstract. Conformance checking allows organizations to verify whether their IT system complies with the prescribed behavior by comparing process executions recorded by the IT system against a process model (representing the normative behavior). However, most of the existing techniques are only able to identify low-level deviations, which provide a scarce support to investigate what actually happened when a process execution deviates from the specification. In this work, we introduce an approach to extract recurrent deviations from historical logging data and generate anomalous patterns representing high-level deviations. These patterns provide analysts with a valuable aid for investigating nonconforming behaviors; moreover, they can be exploited to detect high-level deviations during conformance checking. To identify anomalous behaviors from historical logging data, we apply frequent subgraph mining techniques together with an ad-hoc conformance checking technique. Anomalous patterns are then derived by applying frequent items algorithms to determine highly-correlated deviations, among which ordering relations are inferred. The approach has been validated by means of a set of experiments.

1 Introduction

Organizations are required to monitor their business processes to ensure that their system complies with the prescribed behavior. To this end, organizations usually employ logging mechanisms to record process executions in logs and auditing mechanisms to analyze those logs. Conformance checking has been proposed to assist organizations in verifying whether the observed behavior recorded in an event log matches the prescribed behavior represented as a process model. The notion of *alignment* [1] provides a robust approach to conformance checking, which pinpoints the causes of nonconformity. Given a trace, i.e. a sequence of events recording a process execution, and a process model, an alignment maps the trace to a complete run of the model (see [1] for a formal definition of alignment). Take, for example, a trace $\sigma_1 =$

© Springer International Publishing AG 2017
A. Appice et al. (Eds.): NFMCP 2016, LNAI 10312, pp. 181–197, 2017.
DOI: 10.1007/978-3-319-61461-8_12

⟨customer identification, prepare loan application, check financial status, check external credit rating, check credit purpose, refuse loan⟩ and the loan process in Fig. 1, modeled in the form of a *Petri net* (Definition 1). Figure 2 shows two possible alignments of σ_1 and the net, where activities are abbreviated according to their initial letter(s). The top row of the alignments shows the sequence of events in the trace; the bottom row shows the sequence of activities in the run of the net. Deviations are explicitly shown by columns that contain ≫. For example, the fifth column in γ_1 shows that an activity must occur in σ_1 according to the net, but it is absent in the trace, i.e. a so-called *move on model*. The fourth and fifth columns in γ_2 show that some events occur in the trace although they are not allowed according to the net, i.e. a so-called *move on log*. Other columns for which events in the trace match the activities in the run of the net represent *synchronous moves*.

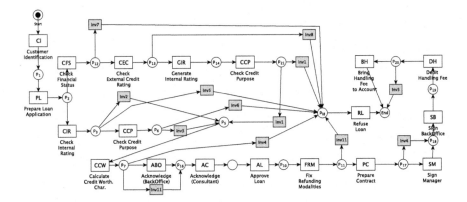

Fig. 1. Loan process represented as a Petri net. Boxes represent transitions (denoting process activities) and circles represent places. The text below the transitions represents the activity label, which is shortened as indicated inside the transitions. Gray boxes represent invisible transitions (i.e. transitions that are not recorded in event logs).

$$\gamma_1 = \frac{|CI|PL|CFS|CEC|\gg|CCP|RL|}{|CI|PL|CFS|CEC|GIR|CCP|RL|} \qquad \gamma_2 = \frac{|CI|PL|\gg|CFS|CEC|CCP|RL|}{|CI|PL|CIR|\gg|\gg|CCP|RL|}$$

Fig. 2. Alignments of $\sigma_1 = \langle CI, PL, CFS, CEC, CCP, RL \rangle$ and the Petri net in Fig. 1

As shown in Fig. 2, there can be several (possibly an infinite number of) alignments of a trace and a Petri net, each of them representing a possible explanation of nonconformity. To determine the quality of alignments, a cost is assigned to each move in the alignment. An optimal alignment of a trace and a Petri net according to a given cost function is the one with the least total cost.

For instance, if we assign cost 1 to moves on log/model and 0 to synchronous moves, γ_1 is the optimal alignment.

Alignments provide diagnostics in terms of *low level deviations*, i.e. elementary deviations like insertions (i.e., moves on log) and suppressions (i.e., moves on model). While low level deviations indicate where the process deviates, they may not provide meaningful diagnostics. Low level deviations need to be analyzed and correlated together into *high level deviations*, e.g. to show whether an activity has been executed instead of another activity or whether the execution of two activities has been swapped.

However, identifying low level deviations and then using them to diagnose high level deviations has a number of drawbacks. First, it requires analysts reexamining the detected deviations to reconstruct what happened, thus resulting in high operational costs. More importantly, it can lead to inaccurate diagnostics. We illustrate this using the alignments in Fig. 2: γ_1 indicates that activity generate internal rating should have been executed, whereas γ_2 indicates that activity check internal rating should have been executed and that check financial status and check external credit rating should not have been executed. The low level deviations in γ_2 can be "interpreted" as a high level deviation indicating that check financial status and check external credit rating were executed instead of check internal rating. An analyst can deem this deviation possible and more plausible than a suppression of generate internal rating (i.e., the analyst would choose this replacement as the explanation of nonconformity rather than to the suppression of generate internal rating). As the number of possible alignments can be infinite, existing alignment-based techniques usually return only optimal alignments, i.e. γ_1 in our case. This alignment, however, does not allow the analyst to reconstruct the deemed deviation. The main problem is that optimal alignments are 'optimal' with respect to low level deviations, and it may not be possible to infer what really happened from the moves in these alignments.

To obtain accurate diagnostics, high level deviations should be treated as 'first class citizens' within conformance checking. Adriansyah et al. [3] show how alignment-based techniques can be adapted to explicitly capture high level deviations using anomalous patterns. Intuitively, an anomalous pattern is an artifact representing a behavior that does not comply with the process model. In particular, Adriansyah et al. construct patterns to detect replacements and swaps of (sequences of) activities as Petri nets and show how these patterns can be used to augment a process model. Existing alignment-based techniques can then be applied to construct alignments that exhibit high-level deviations, providing analysts with accurate diagnostic information. Figure 3a shows an anomalous pattern representing the replacement of activity check internal rating with activities check financial status and check external credit rating, and Fig. 3b shows the alignment of σ_1 and the net of Fig. 1 exhibiting high-level deviations constructed using the approach in [3].

Although the work in [3] makes a first step toward the detection and diagnosis of high level deviations, a number of questions are still left open. In particular:

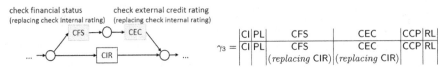

check financial status   check external credit rating
(replacing check internal rating) (replacing check internal rating)

(a) Excerpt of the net in Fig. 1, appended with
the anomalous pattern encoding the replacement.

(b) Alignment of σ_1 and the net in Fig. 1 showing
the replacement.

Fig. 3. Replacement of CIR with $\langle \text{CFS}, \text{CEC} \rangle$

1. *Can we learn patterns representing high level deviations?*

   The approach proposed in [3] is able to identify complex anomalous behaviors for which an anomalous pattern has been defined. In particular, the
   authors provide some predefined patterns to identify replacements and swaps
   of activities. However, we envision that other types of deviations can occur
   in practice. Analysts may want to identify these deviations in their analysis.
   This requires defining patterns capturing the desired anomalous behavior.
   However, the definition of such patterns can be difficult and time consum-
   ing. Thus, it is desirable to provide analysts with tool-supported methods for
   extracting anomalous patterns from past process executions.

2. *Which patterns should be considered in the analysis?*

   An analyst might want to recognize any type of deviation in the constructed
   alignments. However, this significantly increases the search space of alignments, making the approach in [3] unpractical. It is worth noting that some
   anomalous behavior might never occur or be very rare. It is reasonable
   to ignore these deviations, restricting the attention to recurring anomalous
   behaviors which are envisaged to occur in the future.

   In this work, we address these questions. In particular, our goal is to devise
tool-supported methods for the extraction of patterns representing recurrent
(complex) anomalous behaviors from historical logging data. To this end, we
introduce a novel approach to extract partially ordered anomalous subgraphs.
Given an event log and a process model, we apply a frequent subgraph mining
technique to extract relevant subgraphs and propose a conformance checking
algorithm to identify the anomalous ones. Anomalous patterns are derived by
detecting correlated anomalous subgraphs by means of frequent itemset algorithms and inferring ordering relations among them. Our approach for the extraction of anomalous patterns from historical logging data has been validated by
means of a set of experiments.

   The anomalous patterns extracted using our approach can support analysts
in various ways, for instance providing them with an analysis of frequent anomalous behaviors in historical logging data or supporting them in the definition
of a library of anomalous patterns that can be used in combination with conformance checking techniques in the style of [3] to construct alignments exhibiting
high-level deviations as the one shown in Fig. 3b. In the latter application, the
use of the patterns extracted using our approach will allow analysts to obtain

accurate diagnostics concerning recurring anomalous behaviors when analyzing new process executions, relieving them from the burden of reevaluating situations already analyzed.

The remainder of the work is organized as follows. Next section introduces preliminaries on Petri nets. Section 3 details the main steps of the approach. Section 4 presents experimental results. Finally, Sect. 5 discusses related work and Sect. 6 draws some conclusions and delineates future work.

2 Preliminaries

In this work we represent process models using Petri Net notation.

Definition 1 (Labeled Petri Net). *A Labeled Petri net is a tuple* $(P, T, F, A, \ell, m_i, m_f)$ *where* P *is a set of places;* T *is a set of transitions;* $F \subseteq (P \times T) \cup (T \times P)$ *is the flow relation connecting places and transitions;* A *is the set of labels for transitions;* $\ell : T \to A$ *is a function that associates a label with every transition in* T; m_i *is the initial marking;* m_f *is the final marking.*

In a Petri net, transitions represent activities and places represent states. Multiple transitions can have the same label. Such transitions are called duplicate transitions. We distinguish two types of transitions, namely invisible and visible transitions. Visible transitions are labeled with activity names. Invisible transitions are used for routing purposes or related to activities that are not observed by the IT system. Given a Petri net N, the set of activity labels associated with invisible transitions is denoted with $\mathsf{Inv}_N \subseteq A$.

The state of a process model is represented by a marking, i.e. a multiset of tokens on the places of the net. A process model has an initial marking m_i and a final marking m_f. A transition is *enabled* if each of its input places contains at least a token. When an enabled transition is fired (i.e., executed), a token is taken from each of its input places and a token is added to each of its output places.

Process executions are often recorded by the IT system.

Definition 2 (Event, Event Trace, Event Log). *Let* $N = (P, T, F, A, \ell, m_i, m_f)$ *be a labeled Petri net and* $\mathsf{Inv}_N \subseteq A$ *the set of invisible transitions in* N. *An event* e *consists of an executed activity* $a \in A \setminus \mathsf{Inv}_N$. *The set of events is denoted by* $\mathcal{E}$. *An event trace* $\sigma \in \mathcal{E}^*$ *is a sequence of events. An event log* $\mathcal{L} \in \mathbb{B}(\mathcal{E}^*)$ *is a multiset of event traces.*[1]

In this work, we use the coverability graph [23] of a Petri net to assess the conformance of event traces with a process model. Given a Petri net N, the coverability graph of N is a directed graph whose nodes are the markings reachable from the initial marking of N and arcs are labeled by the transitions of N. Intuitively, the coverability graph of a Petri net overapproximates the state space of the net and, thus, exhibits all behaviors allowed by the net. We refer to [23] for a formal definition of coverability graph.

---

[1] $\mathbb{B}(X)$ represents the set of all multisets over X.

3   Methodology

The goal of this work is to support analysts in the detection and analysis of anomalous behaviors in historical logging data. In particular, we present an approach to discover recurrent deviations from the analysis of past process executions and explore their correlation in order to extract anomalous patterns. The extracted patterns can be used, for instance, to augment alignment-based techniques like the one in [3] to construct alignments exhibiting complex anomalous behaviors alongside insertions and suppressions of activities.

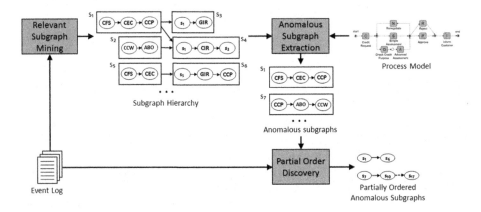

Fig. 4. Overview of the approach

Figure 4 provides an overview of our approach, which comprises three main steps. Given an event log recording past process executions, *relevant subgraphs mining* transforms the log traces into directed graphs and extracts the most relevant subgraphs occurring in the traces using a Frequent Subgraph Mining (FSM) technique. These subgraphs are then analyzed to identify the minimal subgraphs that do not comply with the process model (*anomalous subgraph extraction*). The extracted anomalous subgraphs are used to construct patterns representing frequent (complex) anomalous behavior (*partial order discovery*). In particular, we capture the sets of anomalous subgraphs that frequently occur together by means of frequent itemset discovery algorithms. For each of these itemsets, we infer ordering relations between subgraphs by analyzing their 'position' in the log traces. The obtained structures (i.e., the subgraphs along the partial ordering among them) provide the baseline for the definition of anomalous patterns. In the remainder, we describe each step in detail.

3.1   Relevant Subgraph Mining

The first step of our approach aims to mine relevant subgraphs from the process executions recorded in an event log $\mathcal{L}$. We transform each event trace $\sigma_i \in \mathcal{L}$ into a directed graph $g_i = (V_i, E_i, \phi_i)$, where V_i is the set of nodes, each corresponding to an event in the trace, E_i is the set of the edges, showing ordering relations among the events, and ϕ_i is a labeling function associating each node to the activity of the corresponding event. For the sake of simplicity, in this work we adopt a simple transformation: a node is created for each event in the trace and each pair of subsequent events is linked through an edge. By doing so, every log trace is transformed in a sequence of events ordered according to their order in the log trace. Note that more advanced strategies can be exploited, e.g. to derive graphs also showing possible parallelisms [12]. For instance, many IT systems record both the start and completion time of process activities. Accounting for this information makes it possible to infer situations in which some activities were executed concurrently. We plan to explore methods able to capture and reason on the concurrency occurring in process executions in future work.

To mine relevant subgraphs, we apply a FSM technique, which allows deriving from a given graph set the set of subgraphs whose *support* (i.e., relevance) is above a certain threshold. In this work, we relate the relevance of a subgraph both to its occurrence frequency and size. Given two subgraphs with the same occurrence frequency but different sizes, we are interested in the largest, since we expect to derive a larger amount of knowledge from it. The size of a graph g can be represented in terms of its Description Length (DL), i.e. the number of bits needed to encode its representation (further details on DL are provided in [15]).

To the best of our knowledge, the only FSM algorithm that explicitly considers DL is SUBDUE [18], which evaluates the relevance of a subgraph in terms of its *compression capability*. Namely, given a graph set G and a subgraph s, SUBDUE uses an index based on DL, here denoted by $\nu(s, G)$, which is computed as $\nu(s, G) = \frac{DL(G)}{DL(s) + DL(G|s)}$ where $DL(G)$ is the DL of G, $DL(s)$ is the DL of s and $DL(G|s)$ is the DL of G *compressed* using s, i.e. the graph set in which each occurrence of s is replaced with a single node. The lower is the DL of the compressed dataset, the higher is the compression capability of s.

SUBDUE works iteratively. At each step, it extracts the subgraph with the highest compression capability, which is then used to compress the graph set. The compressed graphs are presented to SUBDUE again. These steps are repeated until no more compression is possible. The outcome of SUBDUE is a hierarchical structure, where mined subgraphs are ordered according to their relevance, showing the existing inclusion relationships among them. Top-level subgraphs are defined only through elements belonging to input graphs (i.e., nodes and arcs), while lower-level subgraphs contain also upper-level subgraphs as nodes. Descending the hierarchy, we pass from subgraphs that are very common in the input graph set to subgraphs occurring in a few input graphs. An example of SUBDUE outcome is shown in Fig. 4. Interested readers can find a description of the approach to extract relevant subgraphs from event logs in [11]. The hierarchical structure of subgraphs discovered by SUBDUE becomes the input of the next step of the methodology.

3.2   Anomalous Subgraph Extraction

The second step aims to extract the subgraphs that do not fit the given process model. To this end, we have developed the *Subgraph Conformance Checking* (SCC) algorithm. In contrast to most of the existing conformance checking algorithms, the SCC algorithm is tailored to check conformance of subgraphs corresponding to portions of process executions. The core idea of the SCC algorithm is to replay a subgraph against the given process model represented by its coverability graph.

Given a subgraph s and the coverability graph R of a Petri net, the SCC algorithm identifies all the arcs in R whose labels match with the first activity of s. Starting from each of these edges, the algorithm checks if there exists a sequence of edges in R whose labels match with the sequence of activities in the subgraph. If such a sequence exists, the subgraph is marked as 'compliant'; otherwise, the subgraph is marked as 'anomalous'. The algorithm is robust with respect to the presence of invisible transitions. Edges labeled with invisible transitions are taken into account while exploring the search space, but are not used while matching the paths with the subgraph. It is easy to observe that checking whether a subgraph corresponds to a path in the coverability graph starting from a given arc, is linear in the length of the subgraph.

Figure 5 shows an example of application of the SCC algorithm. Consider subgraph s_1 in Fig. 5a and (a portion of) the coverability graph corresponding to the Petri net of Fig. 1 in Fig. 5b. The SCC algorithm first looks for arcs labeled with the first event in s_1, i.e. CFS, in the coverability graph. There is only one arc labeled with that activity, i.e. $([p_2], [p_{12}])$. The algorithm marks state $[p_{12}]$ as reachable (denoted by gray) and checks whether there exists an arc outgoing that state with label CEC, i.e. the label of the second node of s_1. As the arc exists (i.e., $([p_{12}], [p_{13}])$), also state $[p_{13}]$ is marked as reachable. From $[p_{13}]$, however, there is not any edge labeled with CCP, i.e. the last event of s_1. Therefore, the subgraph is marked as "anomalous".

Note that if a process involves parallel activities, its coverability graph contains one path for each activities execution ordering allowed by the process model. SCC algorithm can hence be applied also on subgraphs involving parallel activities.

It is worth noting that among the subgraphs mined by SUBDUE there might occur *inclusion* or *overlapping* relationships, i.e. subgraphs can be completely or partially included in other subgraphs. Let $s_i = (V_i, E_i, \phi_i)$ and $s_j = (V_j, E_j, \phi_j)$ be two subgraphs. We say that s_i *includes* s_j, denoted as $s_i \rightarrow_{incl} s_j$, if (i) $\forall v \in V_j$ there exists $v' \in V_i$ s.t. $\phi_i(v) = \phi_j(v')$ and (ii) $\forall (u, v) \in E_j$ there exists $(u', v') \in E_i$ s.t. $\phi_i(u) = \phi_j(u')$ and $\phi_i(v) = \phi_j(v')$. We say that s_i *overlaps* s_j, if exists a subgraph s_z such that s_i strictly includes s_z and s_j strictly includes s_z.

The presence of inclusion relationships affects the extraction of correlations among deviations and, consequently, the analysis of the outcome of the approach. In fact, subgraphs related by an inclusion relationship are highly correlated; however, this information is redundant and only introduces noise into the analysis. Given two subgraphs s_i, s_j such that $s_i \rightarrow_{incl} s_j$, it is easy to observe that

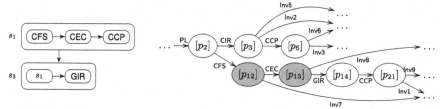

(a) Hierarchical structure of subgraphs to be verified   (b) Portion of the coverability graph of the Petri net in Fig. 1. The states marked in gray are the reachable states.

Fig. 5. Example of application of SCC algorithm.

s_j is able to detect all process executions containing s_i but not the other way around. s_j is hence more general than s_i and thus preferable for the definition of anomalous patterns. To capture this intuition, we introduce the notion of *minimal* anomalous subgraphs. Given a set of subgraphs $S = \{s_1, \ldots, s_n\}$, the set of minimal subgraphs is $S_{min} = \{s_i \mid s_i \in S \land \nexists s_j \in S_{min} \text{ s.t. } s_i \rightarrow_{incl} s_j\}$.

For the definition of anomalous patterns, we only consider minimal subgraphs. Given the hierarchical structure returned by SUBDUE, we start assessing the conformance of the root subgraphs using the SCC algorithm. If a subgraph is marked as 'anomalous', the algorithm prunes all the branches involving the descendants of the subgraph, since, although they are anomalous 'by inheritance', none of them is minimal. Otherwise, if a subgraph fits the process model, it is marked as 'compliant' and its child subgraphs are iteratively analyzed using the SCC algorithm. The algorithm terminates when all subgraphs in the hierarchical structure of subgraphs computed by SUBDUE are marked as either 'complaint' or 'anomalous'. In the example of Fig. 5a, subgraph s_3 is marked as 'anomalous' as it includes an anomalous subgraph (i.e., s_1).

Overlapping subgraphs, on the other hand, can provide useful insights about potential anomalous behavior. In fact, two subgraphs overlap on some activities if there are some process executions which differ before/after those activities. By analyzing overlapping subgraphs that frequently occur together, we can explore some portions of these alternative execution paths. Thus, we consider subgraphs relationships in the final step of the methodology, as explained in the following section.

3.3 Partial Order Discovery

The final step of the approach aims to derive ordering relations among minimal anomalous subgraphs. These ordering relations are used to generate anomalous patterns, i.e. partially ordered subgraphs that show how apparently different anomalous behaviors are usually correlated. First, we generate an occurrence matrix where each cell c_{ij} represents the number of occurrence of the j-th subgraph in the i-th trace. We apply well-known *frequent itemset algorithms* [14] to this matrix, thus deriving all the subgraphs which co-occur with a support above a given threshold. To determine how the subgraphs in a frequent itemset are combined, we infer ordering relations between the elements of the itemset pairwise. More precisely, for each pair of subgraphs s_i, s_j belonging to the same itemset, we define one of the following relations: *(i)* the *sequentially* relation, denoted as $s_i \rightarrow_{seq} s_j$, which states that s_j occurs immediately after s_i, *(ii)* the *overlapping* relation, denoted as $s_i \rightarrow_{ov} s_j$, which states that s_i occurs before s_j and their executions overlap, *(iii)* the *eventually* relation, denoted as $s_i \rightarrow_{ev} s_j$, which states that s_j will occur after s_i, but an arbitrary number of other activities (at least one) occurred between the two subgraphs. To derive these relations, we analyze the position of the events forming each subgraph of the itemset in the log traces in which the itemset occurs. In particular, we evaluate the occurrence frequency of sequentially, overlapping and eventually relations by means of M_{seq}, M_{ov} and M_{ev} matrices respectively. Each cell of a matrix represents the number of times in which the ordering relation represented by the matrix occurred for a given pair of subgraphs. It is worth noting that in the presence of noisy logs we can detect unreliable relations. To deal with this issue, we consider only ordering relations whose occurrence frequency is above a given threshold.

As an example, let consider the frequent itemset $\{s_{26}, s_{266}, s_{67}\}$ where $s_{26} = \langle \mathsf{CI}, \mathsf{CI} \rangle$, $s_{266} = \langle \mathsf{PL}, \mathsf{CCP} \rangle$ and $s_{67} = \langle \mathsf{CCP}, \mathsf{CIR} \rangle$. Analyzing the positions of subgraphs in the log traces in which these subgraphs occur (see Fig. 6a for an example of such traces), we can observe that s_{266} usually occurs immediately after s_{26} (i.e., $s_{26} \rightarrow_{seq} s_{266}$). Moreover, we can observe that s_{266} overlaps s_{67} (i.e., $s_{266} \rightarrow_{ov} s_{67}$) and s_{67} eventually occurs after s_{26} (i.e., $s_{26} \rightarrow_{ev} s_{67}$). Figure 6b shows matrices M_{seq}, M_{ov} and M_{ev} for itemset $\{s_{26}, s_{266}, s_{67}\}$. As can be observed in the matrices, these relations are reliable as they have a high occurrence frequency. Figure 6c shows the obtained partially ordered subgraph.

(a) Excerpt of traces including subgraphs s_{26}, s_{266} and s_{67}

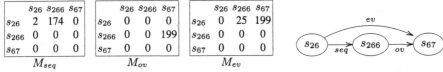

(b) Ordering Relations Matrices   (c) Partially Ordered Subgraph

Fig. 6. Ordering relations discovery for itemset $\{s_{26}, s_{266}, s_{67}\}$.

4   Experiments

We have implemented our approach as two modules of the ESub tool [10], namely *Anomalous Subgraphs Checking* (implementing steps 1 and 2) and *Partial Order*

Discovery (implementing step 3).[2] The first module takes as input an event log and the coverability graph of a Petri net and uses SUBDUE to generate a hierarchical structure of subgraphs and the SCC algorithm to extract the anomalous subgraphs. Figure 7a shows a screenshot of the module displaying a portion of the hierarchical structure derived by SUBDUE where anomalous subgraphs are denoted by a (red) thick border, their children by a (orange) dotted border and compliant subgraphs by a (green) normal border. The second module takes as input the set of frequent itemsets and the graphs generated from the log and derives the partially ordered subgraphs. A screenshot of this module is shown in Fig. 7b. Each edge is labeled with the type of relation it represents. Normal lines are used for sequentially relations, bold lines for overlapping relations and dotted lines for eventually relations.

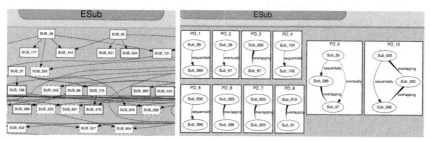

(a) Anomalous Subgraphs Checking   (b) Partial Order Discovery

Fig. 7. ESub modules for anomalous pattern extraction. (Color figure online)

To evaluate the approach we performed a number of experiments using a synthetic event log generated by simulating the Petri net in Fig. 1. This model represents a real-world loan application management process, which has been defined and validated through interviews with the managers of a bank [2]. The results discussed in this section can hence be considered, to a certain extent, representative of the outcome that can be obtained in real-world contexts. Based on the process model in Fig. 1, we generated 3905 traces consisting of 43673 events using CPN Tools (http://cpntools.org/).

To encompass anomalous behaviors in the generated event log, we artificially manipulated the log by introducing noise. In particular, we introduced 20% of noise to each generated event trace by adding or removing some events, randomly chosen among the activities of the model. In addition, we inserted some high-level deviations, namely *swaps*, *repetitions* and *replacements*. A swap occurs when two or more activities are executed in the opposite order compared to the order defined by the model; we swapped the execution of sequence ⟨CCP⟩ with the one of ⟨CIR⟩ in 18.0% of the traces, and the execution of ⟨GIR⟩ with the execution of ⟨CFS⟩ in 15.5% of the traces. A repetition means that a given (sequence of)

---

[2] http://kdmg.dii.univpm.it/?q=content/esub.

activity(ies) is repeated multiple times (without belonging to a loop); we added two repetitions, namely the repetition of sequence ⟨CI⟩ and of sequence ⟨CER⟩ in 33.6% and 9.0% of the traces respectively. Finally, a replacement indicates that a given (sequence of) activity(ies) is executed instead of another one; in our experiments, sequence of activities ⟨ACO⟩ was replaced with sequence ⟨SBO⟩, ⟨ACO, AL⟩ with ⟨PLA, BHF, GIR⟩ and ⟨PC, S⟩ with ⟨CCP, CCP⟩ in 12.0%, 3.7% and 12.0% of the traces respectively.

By doing so, we obtain an event log involving several heterogeneous anomalous behaviors, among which is however possible to recognize some regularities. This reflects what we reasonable expect to find in a real-world context. Note that this implies that we cannot expect to detect patterns with a very high support value. However, this does not affect the validity of the approach. The importance of a deviation is not necessarily related to its frequency. For instance, in several domains (e.g., security), undesired behaviors have to be detected even if they are not very frequent.

The event log was given as input to the Anomalous Subgraphs Checking module. SUBDUE extracted 1245 subgraphs, from which 186 minimal anomalous subgraphs were identified using the SCC algorithm. This set of minimal anomalous subgraphs was used to derive the frequent itemsets. For our experiments, we used FP-Growth [14] with a minimum support threshold of 5%. The outcome of the FP-Growth algorithm was passed to the Partial Order Discovery module. Based on the derived itemsets and a threshold of 40%, the module extracted ten partially ordered subgraphs. Table 1 shows the support of each pattern with respect to the traces where its itemset occurs (δ_{item}) and with respect to the overall set of traces (δ_{all}). We can observe that ordering relations hold for most of the occurrences of their itemsets. Moreover, most of the derived partial orders have good support values, higher than (or anyway closed to) 5%. Note that the support of a pattern is always lower or at most equal to the support of its corresponding itemset. In fact, there can be some traces where the itemset occurs, but its subgraphs do not match the ordering relations of the pattern. This explains why po_1 and po_9 have a δ_{all} lower than 5%.

Table 1. Support values of the discovered partially ordered subgraphs with respect to the traces where the itemset occurs (δ_{item}) and with respect to all traces (δ_{all}).

Id	po_1	po_2	po_3	po_4	po_5	po_6	po_7	po_8	po_9	po_{10}
δ_{item}	85.6	96.7	99.8	97.3	96.8	100	99.6	96.4	86.4	99.5
δ_{all}	4.7	5.2	17.9	14.7	11.6	12.2	11.9	8.3	4.4	11.6

We observed that all high-level deviations that were inserted during the generation of the event log are captured by the discovered partially ordered subgraphs. Moreover, the support values of those patterns are coherent with the support of the corresponding deviations. Therefore, the results demonstrate that our approach is able to detect frequent anomalous behaviors. Note that some of the discovered patterns encompass the combination of the inserted high-level deviations with low-level deviations (i.e., inserted and skipped activities) randomly inserted in log traces and/or with other high-level deviations, thus originating behaviors more complex than the designed ones.

To provide a concrete example of the outcome of the approach and illustrate its capability, next we discuss in detail some of the discovered patterns. For the sake of space, we only focus on three of them, namely po_4, po_5 and po_9, which allow us to point out some interesting aspects of the approach.

Partially ordered subgraphs po_4 (Fig. 8a) and po_5 (Fig. 9) show a swap of activities CFS and GIR and a replacement of sequence $\langle PC, SB \rangle$ with sequence $\langle CCP, CCP \rangle$, respectively. The support of these patterns (see Table 1) is close with the support we set for the deviations: the support of po_4 is 14.7 whereas the execution of CFS and GIR was swapped in 15.5% of the traces; the support of po_5 is 11.6, whereas sequence $\langle PC, SB \rangle$ was replaced by sequence $\langle CCP, CCP \rangle$ in 12% of the traces. The difference is due to some traces which do not fit the ordering relations of the patterns because of other inserted/deleted activities. We would like to point out that detecting these high-level deviations is quite straightforward by analyzing po_4 and po_5. On the other hand, detecting them using, for instance, alignments requires additional efforts by the human analyst and is usually far from trivial. For example, Fig. 8b shows the alignments returned by the ProM plug-in *PNetReplayer* for the trace $\sigma_5 = \langle CI, CI, PL, GIR, CEC, CFS, CCP, RL \rangle$ and the net of Fig. 1, which involves the swap of po_4, and Fig. 9b shows the alignments for trace $\sigma_{13} = \langle CI, CI, PL, CIR, CCP, CCW, AC, AL, FRM, CCP, CCP, DH, BH \rangle$, which involves the replacement exhibited by po_5. Using alignments, the analyst has to relate several not synchronous moves to recognize the occurrence of a high-level deviation. For instance, in order to derive the replacement from γ_5 (Fig. 9b) the analyst has to relate the deletion of activities PC and SB with the insertion of activities CCP and CCP. The detection of the swap in γ_4 (Fig. 8b) requires even more efforts, since requires to relate the deletion of CFS and the insertion of GIR, occurring before CEC, with the deletion of GIR and the insertion of CFS occurring after CEC. Clearly, this can easily lead to misleading diagnostics, especially when more than one activity occur between the activities involved by the high-level deviation or other deviations occur.

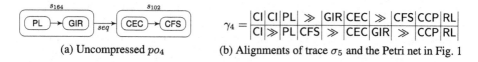

(a) Uncompressed po_4

$$\gamma_4 = \frac{|CI|CI|PL| \gg |GIR|CEC| \gg |CFS|CCP|RL|}{|CI| \gg |PL|CFS| \gg |CEC|GIR| \gg |CCP|RL|}$$

(b) Alignments of trace σ_5 and the Petri net in Fig. 1

Fig. 8. Analysis of anomalous pattern po_4

Partially ordered subgraph po_9 (Fig. 6c) corresponds to the combination of two high-level deviations, namely the repetition of activity CI and the swap of activities CCP and CIR, thus originating a new, not a-priori known, pattern. This provides an example of the capability of our approach to extract general

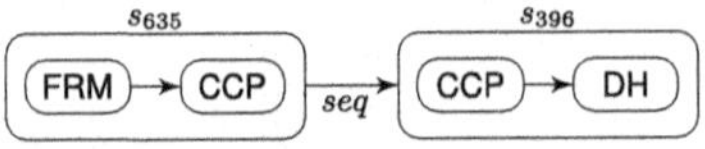

(a) Uncompressed po_5

$$\gamma_5 = \frac{|CI|CI|PL|CIR|CCP|CCW|AC|AL|FRM|\gg|\gg|CCP|CCP|DH|BH|}{|CI|\gg|PL|CIR|CCP|CCW|AC|AL|FRM|PC|SB|\gg|\;|\gg|DH|BH|}$$

(b) Alignments of trace σ_{13} and the Petri net in Fig. 1

Fig. 9. Analysis of anomalous pattern po_5

patterns that do not necessarily reflect a-priori knowledge of deviations. On the other hand, it also points out the need of post-processing the discovered patterns; it is easy to see that the edge between s_{26} and s_{67} is not needed to interpret the pattern. We plan to address this issue in future work.

5   Related Work

A number of approaches have been proposed for conformance checking. Some approaches [7,9,22] check whether event traces satisfy a set of compliance rules. Rozinat and van der Aalst [21] propose a token-based technique to replay event traces over a process model and use the information obtained from remaining and missing tokens to detect deviations. Banescu et al. [6] extend the work in [21] to identify and classify high level deviations by analyzing the configuration of remaining and missing tokens. However, it has been shown that token-based techniques can provide misleading diagnostics.

Recently, alignments have been proposed as a robust approach to conformance checking [1]. Alignments are able to pinpoint deviations causing nonconformity based on a given cost function. These cost functions, however, are usually based on human judgment and, hence, prone to imperfections, which can ultimately lead to incorrect diagnostics. To obtain probable explanations of nonconformity, Alizadeh et al. [4] propose an approach to compute the cost function by analyzing historical logging data, which is extended in [5] to consider multiple process perspectives. Alignment-based techniques rely on total ordering of events; thus, diagnostics obtained by these techniques can be unreliable when timestamps of events are coarse or incorrect. Lu et al. [20] describe how partially ordered traces can be obtained from sequential event logs and propose an approach for computing partially ordered alignments using these partially ordered traces. However, alignment-based approaches usually provide the diagnostic information in terms of low level deviations. Our work, instead, focuses on the identification and analysis of high-level deviations. Adriansyah et al. [3] show how alignment-based techniques can be extended to directly capture high level deviations in alignments using (a few simple) predefined anomalous patterns. Our approach complements the work in [3] by showing how to construct anomalous patterns from the analysis of historical logging data. In particular, our approach constructs anomalous patterns by extracting recurrent nonconforming subprocesses from event logs.

Several approaches for subprocess extraction have been proposed in the literature. Well-known approaches for the extraction of subprocesses from sequential traces are: [8], which detects subprocesses by identifying sequences of events that fit a-priori defined templates; [16], which exploits a sequence pattern mining algorithm to derive frequent sequences of clinical activities from clinical logs; and [19], which introduces an approach to derive "episodes", i.e. directed graphs where nodes correspond to activities and edges to *eventually-follow* precedence relations, which, given a pair of activities, state which one occurs later. Compared to these approaches, the one proposed in this work does not require defining any predefined template and extracts the subprocesses that are the most relevant according to their description length, thus taking into account both frequency and size in determining the relevance of each subprocess. Other approaches aim to convert traces into directed graphs representing execution flows and, then, apply frequent subgraph mining techniques to derive the most relevant subgraphs. For instance, Hwang et al. [17] generates "temporal graphs", where two nodes are linked only if the corresponding activities have to be executed sequentially. The applicability of this approach, however, is limited to event logs storing starting and completion time of events. Greco et al. [13] proposes a FSM algorithm that exploits knowledge about relationships among activities (e.g., AND/OR split) to drive subgraphs mining. Graphs are generated by replaying traces over the process model; however, this algorithm requires a model properly representing the event log, which may not be available for many real-world processes. In contrast, our approach for subprocess extraction does not require neither the presence of special attributes in the event log nor a-priori models of the process or other domain knowledge.

6   Conclusions and Future Work

In this work, we have presented a novel approach to discover complex anomalous patterns from historical logging data, showing high-level deviations in process executions. Main novelties consists in *(i)* an approach to extract anomalous subgraphs representing raw deviations and *(ii)* an approach to derive partially ordered anomalous subgraphs representing complex anomalous behaviors. Our experiments demonstrated the capability of the approach by returning meaningful patterns capturing high-level deviations that, on the other hand, would be hard to identify using, e.g. alignment-based techniques.

Although the experiments show promising results, more efforts are required in order to move from partially ordered anomalous subgraphs, describing basic ordering relations, to anomalous subprocesses describing the execution flows of deviations. First, a post-processing of the discovered patterns is needed to remove redundant relations, as mentioned in Sect. 4. Moreover, it is desirable to derive more complex flow constructs, e.g. loops and AND/OR relations. A possible direction in this regard consists in investigating the application of process discovery algorithms. A further extension consists in devising (semi)automatic techniques able to detect in which portions of the process anomalous subprocesses occurred, thus simplifying the analysis of deviations. Extending the original model with the detected subprocesses also paves the way for implementing efficient strategies to detect future instances of anomalous behaviors, both in an on-line and an off-line setting. This can be obtained by investigating how to combine our approach with the one proposed in [3].

In future work, we plan to address these issues. Furthermore, we intend to perform more extensive experiments on real-life event logs, exploring also other approaches, for instance, model building approaches.

Acknowledgment. This work has been partially funded by the NWO CyberSecurity programme under the PriCE project and by the Dutch national program COMMIT under the THeCS project.

References

1. van der Aalst, W., Adriansyah, A., van Dongen, B.: Replaying history on process models for conformance checking and performance analysis. Wiley Int. Rev. Data Min. Knowl. Discov. **2**(2), 182–192 (2012)
2. Accorsi, R., Stocker, T.: On the exploitation of process mining for security audits: the conformance checking case. In: Proceedings of Annual Symposium on Applied Computing, pp. 1709–1716. ACM (2012)
3. Adriansyah, A., van Dongen, B.F., Zannone, N.: Controlling break-the-glass through alignment. In: Proceedings of International Conference on Social Computing, pp. 606–611. IEEE (2013)
4. Alizadeh, M., de Leoni, M., Zannone, N.: History-based construction of alignments for conformance checking: formalization and implementation. In: Ceravolo, P., Russo, B., Accorsi, R. (eds.) SIMPDA 2014. LNBIP, vol. 237, pp. 58–78. Springer, Cham (2015). doi:10.1007/978-3-319-27243-6_3
5. Alizadeh, M., de Leoni, M., Zannone, N.: Constructing probable explanations of nonconformity: a data-aware and history-based approach. In: Proceedings of Symposium Series on Computational Intelligence, pp. 1358–1365. IEEE (2015)
6. Banescu, S., Petković, M., Zannone, N.: Measuring privacy compliance using fitness metrics. In: Barros, A., Gal, A., Kindler, E. (eds.) BPM 2012. LNCS, vol. 7481, pp. 114–119. Springer, Heidelberg (2012). doi:10.1007/978-3-642-32885-5_8
7. Borrego, D., Barba, I.: Conformance checking and diagnosis for declarative business process models in data-aware scenarios. Expert Syst. Appl. **41**(11), 5340–5352 (2014)
8. Jagadeesh Chandra Bose, R.P., van der Aalst, W.M.P.: Abstractions in process mining: a taxonomy of patterns. In: Dayal, U., Eder, J., Koehler, J., Reijers, H.A. (eds.) BPM 2009. LNCS, vol. 5701, pp. 159–175. Springer, Heidelberg (2009). doi:10.1007/978-3-642-03848-8_12
9. Caron, F., Vanthienen, J., Baesens, B.: Comprehensive rule-based compliance checking and risk management with process mining. Decis. Support Syst. **54**(3), 1357–1369 (2013)

10. Diamantini, C., Genga, L., Potena, D.: Esub: exploration of subgraphs. In: Proceedings of the BPM Demo Session, pp. 70–74 (2015). CEUR-WS.org
11. Diamantini, C., Genga, L., Potena, D.: Behavioral process mining for unstructured processes. J. Intell. Inf. Syst. **47**(1), 5–32 (2016)
12. Diamantini, C., Genga, L., Potena, D., van der Aalst, W.: Building instance graphs for highly variable processes. Expert Syst. Appl. **59**, 101–118 (2016)
13. Greco, G., Guzzo, A., Manco, G., Saccà, D.: Mining and reasoning on workflows. IEEE Trans. Knowl. Data Eng. **17**(4), 519–534 (2005)
14. Han, J., Pei, J., Yin, Y.: Mining frequent patterns without candidate generation. In: ACM Sigmod Record, vol. 29, pp. 1–12. ACM (2000)
15. Holder, L., Cook, D., Djoko, S.: Substructure discovery in the SUBDUE system. In: Proceedings of the AAAI Workshop on Knowledge Discovery in Databases, pp. 169–180 (1994)
16. Huang, Z., Lu, X., Duan, H.: On mining clinical pathway patterns from medical behaviors. Artif. Intell. Med. **56**(1), 35–50 (2012)
17. Hwang, S., Wei, C., Yang, W.: Discovery of temporal patterns from process instances. Comput. Indus. **53**(3), 345–364 (2004)
18. Jonyer, I., Cook, D., Holder, L.: Graph-based hierarchical conceptual clustering. J. Mach. Learn. Res. **2**, 19–43 (2002)
19. Leemans, M., Aalst, W.M.P.: Discovery of frequent episodes in event logs. In: Ceravolo, P., Russo, B., Accorsi, R. (eds.) SIMPDA 2014. LNBIP, vol. 237, pp. 1–31. Springer, Cham (2015). doi:10.1007/978-3-319-27243-6_1. CEUR-ws.org
20. Lu, X., Fahland, D., Aalst, W.M.P.: Conformance checking based on partially ordered event data. In: Fournier, F., Mendling, J. (eds.) BPM 2014. LNBIP, vol. 202, pp. 75–88. Springer, Cham (2015). doi:10.1007/978-3-319-15895-2_7
21. Rozinat, A., van der Aalst, W.M.P.: Conformance checking of processes based on monitoring real behavior. Inf. Syst. **33**(1), 64–95 (2008)
22. Taghiabadi, E.R., Gromov, V., Fahland, D., der Aalst, W.M.P.: Compliance checking of data-aware and resource-aware compliance requirements. In: Meersman, R., Panetto, H., Dillon, T., Missikoff, M., Liu, L., Pastor, O., Cuzzocrea, A., Sellis, T. (eds.) OTM 2014. LNCS, vol. 8841, pp. 237–257. Springer, Heidelberg (2014). doi:10.1007/978-3-662-45563-0_14
23. Valk, R., Vidal-Naquet, G.: Petri nets and regular languages. J. Comput. Syst. Sci. **23**(3), 299–325 (1981)

Mining Spatio-Temporal Patterns of Periodic Changes in Climate Data

Corrado Loglisci[1,2(✉)], Michelangelo Ceci[1,2], Angelo Impedovo[1,2], and Donato Malerba[1,2]

[1] Department of Computer Science, Universita' degli Studi di Bari "Aldo Moro", Bari, Italy
{corrado.loglisci,michelangelo.ceci,angelo.impedovo, donato.malerba}@uniba.it
[2] CINI - Consorzio Interuniversitario Nazionale per l'Informatica, Rome, Italy

Abstract. The climate changes have attracted always interest because they may have great impact on the life on Earth and living beings. Computational solutions may be useful both for the prediction of the climate changes and for their characterization, perhaps in association with other phenomena. Due to the cyclic and seasonal nature of many climate processes, studying their repeatability may be relevant and, in many cases, determinant. In this paper, we investigate the task of determining changes of the weather conditions, which are periodically repeated over time and space. We introduce the spatio-temporal patterns of periodic changes and propose a computational solution to discover them. These patterns allows us to represent spatial regions with same periodic changes. The method works on a grid-based data representation and relies on a time-windows analysis model to detect periodic changes in the grid cells. Then, the cells with same changes are selected to form a spatial region of interest. The usefulness of the method is demonstrated on a real-world dataset collecting weather conditions.

1 Introduction

Climatology is a discipline essentially focused on the study of the weather conditions and it is one of the scientific fields characterized by a large variety of data-intensive and dynamic processes. Studying the evolution of the weather becomes thus determinant because might support the understanding of other processes, such as the industrialization and atmospheric changes. In this sense, a valid contribution is represented from the application of data-driven techniques [5], which opens to the possibility to analyze climate observations in order to unearth empirical knowledge without demanding a-priori hypothesis, as the standard statistics method do instead. The proliferation of the technologies able to record and store massive meteorological data has definitely confirmed the usefulness of the data analysis algorithms for several problems in Climatology.

One of the most scientifically and technologically challenging problems is building and refining predictive models with changes and events of the weather

© Springer International Publishing AG 2017
A. Appice et al. (Eds.): NFMCP 2016, LNAI 10312, pp. 198–212, 2017.
DOI: 10.1007/978-3-319-61461-8_13

conditions. Although in data mining we can find a long list of works on event and change detection [3], the identification of changes in climate data is challenging for several reasons. First, climate data tend to be noisy, therefore we could have difficulty in distinguishing, with an high degree of certainty, the difference between significant changes and spurious outliers. Second, changes that persist over time and that cover relatively long intervals of time (e.g., days) can be originated from instantaneous deviations (e.g., rainfall extreme events which span few hours), which we could erroneously assess as meaningless. Third, the global models provide reliable indications for world-wide climate, while they could be no longer appropriate capture features of the regional weather conditions, where instead local models could be effective [17].

In Climatology, many phenomena are cyclic in nature and can exhibit repetitive behaviors. Likewise, changes in weather conditions can be periodic because they can be repeated at regular intervals of time. For instance, seasonal changes reflect the occurrence of the expected variations of the weather conditions and can recur up to one year of distance. The periodicity becomes thus a good indicator of the repeatability and meaningfulness of the changes since the variations which regularly recur may be considered more interesting than those episodic.

This paper focuses on the analysis of time-series describing the weather conditions recorded in geographically distributed locations and, in particular, introduces the problem to discover spatio-temporal patterns able to relate periodic changes of the weather conditions with the spatial regions in which the changes occur. The geographic information of the weather conditions is used to determine the spatial component of the patterns, while the periodicity associated with the changes denotes the temporal component of the patterns. In this work, we propose a data mining framework which analyzes weather conditions data partitioned over a gridded data space. It proceeds in two subsequent steps, first detects periodic changes at the level of individual cells of the grid and then it finds sequential patterns of the periodic changes only over the cells in which the changes are present. The use of a technique of data partitioning is to not under-estimate the periodicity of local changes, which instead we could experience working on (global) statistical regularities. More precisely, in the first step, we combine a time windows-based analysis model with a frequent pattern mining method, in order to search for periodic changes in each grid cell. Changes are detected as significant variations of the frequency of the patterns mined from two different time-windows of data. The rationale in using the frequency is that it denotes regularity, therefore frequent patterns can provide empirical evidence about changes really happened. Building time-windows allows us to summarize the changes occurring at the level of time instants and model them at a higher level of temporal granularity, that is, intervals of time. Not all the changes are considered, but only those which are repeated over time-windows in several grid cells. The second step operates on the detected periodic changes and uses a sequential pattern mining method, in order to find changes common to different cells. Sequential patterns allows us to find changes at a higher level of spatial granularity based on aggregations of cells.

The paper is organized as follows. In Sect. 2, we report necessary notions, while the method is described in Sect. 3. An application to the real-world dataset is described in Sect. 4. Then, we discuss the related literature (Sect. 5). Finally, conclusions close the paper (Sect. 6).

2 Basics and Definitions

Before formally describing the proposed method, we report basic notions and definitions necessary for the paper.

Let $\{t_1 \ldots t_n\}$ be a sequence of discrete time-points. For each time-point t_i, we have the values $A_i \in \Re^d$ of the weather parameters measured in geographically distributed areal units. A *time-window* τ is a sequence of consecutive time-points $\{t_i, \ldots, t_j\}$ ($t_1 \leq t_i, t_j \leq t_n$), which we denote as $[t_i; t_j]$. The width w of a time-window is the number of time-points in τ, i.e. $w = j - i + 1$. We assume that all the time-windows have the same width w. Two time-windows τ and τ' defined as $\tau = [t_i; t_{i+w-1}]$ and $\tau' = [t_{i+w}; t_{i+2w-1}]$ are *consecutive*.

Let $\tau = [t_i; t_{i+w-1}]$, $\tau' = [t_{i+w}; t_{i+2w-1}]$, $\tau'' = [t_j; t_{j+w-1}]$, and $\tau''' = [t_{j+w}; t_{j+2w-1}]$ be time-windows, two pairs of consecutive time-windows (τ, τ') and (τ'', τ''') are δ-*separated* if $(j + w) - (i + w) \leq \delta$ ($\delta > 0, \delta \geq w$). Two pairs of consecutive time-windows (τ, τ') and (τ'', τ''') are *chronologically ordered* if j > i. In the remaining of the paper, we use the notation τ_{h_k} to refer to a time-window and the notation (τ_{h_1}, τ_{h_2}) to indicate a pair of consecutive time-windows.

The following notions are crucial for this work. A pattern P is a set of pairs, each pair is composed by a weather parameter and its value. It can have at most d pairs, which is the number of weather parameters. We say that P occurs at a time-point t_i if all pairs of P occur at the same time-point t_i. A pattern P is characterized by a statistical parameter, namely the *support* (denoted as $sup_{\tau_{h_k}}(P)$), which denotes the relative frequency of P in the time-window τ_{h_k}. It is computed as the number of the time-points of τ_{h_k} in which P occurs divided by the total number of time-points of τ_{h_k}. When the support exceeds a minimum user-defined threshold $minSUP$, P is *frequent* (FP) in the time-window τ_{h_k}.

Definition 1. Emerging Pattern (EP)
Let (τ_{h_1}, τ_{h_2}) be a pair of consecutive time-windows; P be a frequent pattern in the time-windows τ_{h_1} and τ_{h_2}; $sup_{\tau_{h_1}}(P)$ and $sup_{\tau_{h_2}}(P)$ be the support of the pattern P in τ_{h_1} and τ_{h_2} respectively, P is an emerging pattern in (τ_{h_1}, τ_{h_2}) iff
$$\frac{sup_{\tau_{h_1}}(P)}{sup_{\tau_{h_2}}(P)} \geq minGR \vee \frac{sup_{\tau_{h_2}}(P)}{sup_{\tau_{h_1}}(P)} \geq minGR$$

where, $minGR$ (>1) is a user-defined minimum threshold.
The ratio $sup_{\tau_{h_1}}(P)/sup_{\tau_{h_2}}(P)$ $(sup_{\tau_{h_2}}(P)/sup_{\tau_{h_1}}(P))$ is denoted with $GR_{\tau_{h_1}, \tau_{h_2}}(P)$ $(GR_{\tau_{h_2}, \tau_{h_1}}(P))$ and it is called *growth-rate* of P from τ_{h_1} to τ_{h_2} (from τ_{h_2} to τ_{h_1}). When $GR_{\tau_{h_1}, \tau_{h_2}}(P)$ exceeds $minGR$, the support of P decreases from τ_{h_1} to τ_{h_2} by a factor equal to the ratio $sup_{\tau_{h_1}}(P)/sup_{\tau_{h_2}}(P)$, while when $GR_{\tau_{h_2}, \tau_{h_1}}(P)$ exceeds $minGR$, the support of P increases by a factor equal to $sup_{\tau_{h_2}}(P)/sup_{\tau_{h_1}}(P)$.

The concept of emerging pattern is not novel in the literature [4]. In its classical formulation, it refers to the values of support of a pattern discovered on two different classes of data, while, in this work, we extend that notion to represent the differences between the data collected in two intervals of time, and therefore, we refer to the values of support of a pattern which has been discovered on two time-windows.

Definition 2. Periodic Change (PC)

Let T : $\langle(\tau_{i_1}, \tau_{i_2}), \dots, (\tau_{m_1}, \tau_{m_2})\rangle$ be a sequence of chronologically ordered pairs of time-windows; P be an emerging pattern between the time-windows τ_{h_1} and $\tau_{h_2}, \forall h \in \{i, \dots, m\}$; $\langle GR_{\tau_{i_1}, \tau_{i_2}}, \dots, GR_{\tau_{m_1}, \tau_{m_2}}\rangle$ be the values of growth-rate of P in the pairs $\langle(\tau_{i_1}, \tau_{i_2}), \dots, (\tau_{m_1}, \tau_{m_2})\rangle$ respectively; $\Theta_P : \Re \to \Psi$ be a function which maps $GR_{\tau_{h_1}, \tau_{h_2}}(P)$ to a nominal value $\psi_{\tau_{h_1}, \tau_{h_2}} \in \Psi, \forall h \in \{i, \dots, m\}, P$ is a periodic change iff:

1. $|T| \geq minREP$
2. (τ_{h_1}, τ_{h_2}) and (τ_{k_1}, τ_{k_2}) are δ-separated $\forall h \in \{i, \dots, m-1\}$, $k = h+1$ and there is no pair $(\tau_{l_1}, \tau_{l_2}), h < l$, s.t. (τ_{h_1}, τ_{h_2}) and (τ_{l_1}, τ_{l_2}) are δ-separated
3. $\psi = \psi_{\tau_{i_1}, \tau_{i_2}} = \dots = \psi_{\tau_{m_1}, \tau_{m_2}}$

where, $minREP$ is a minimum user-defined threshold. The function Θ is used to handle the numerical information associated to the growth-rate and allows us to crisply distinguish the magnitude of different growth-rate values. A PC is a frequent pattern whose support increases (decreases) at least $minREP$ times with an order of magnitude greater than $minGR$. Each change (increase/decrease) occurs within δ time-points and it is represented by the nominal value $\psi \in \Psi$. We denote a periodic change PC with the notation $\langle P, T, \psi\rangle$. An example of periodic change is reported here. Consider the pattern

$$P : air\_temperature = [301; 307], pressure = [95; 100], relative\_humidity = [60; 70]$$

where $sup_{apr\_2011}(P) = sup_{apr\_2012}(P) = sup_{apr\_2013}(P) = 0.25, sup_{may\_2011}(P) = sup_{may\_2012}(P) = sup_{may\_2013}(P) = 0.5, sup_{nov\_2011}(P) = sup_{nov\_2012}(P) = sup_{nov\_2013}(P) = 0.5, sup_{dec\_2011}(P) = sup_{dec\_2012}(P) = sup_{dec\_2013}(P) = 0.1$. Here, the values of the support of the pattern P increase through the pairs of the windows $[apr\_2011, may\_2011]$, $[apr\_2012, may\_2012]$ and $[apr\_2013, may\_2013]$ respectively, indeed the values of growth-rate $GR_{apr\_2011, may\_2011}(P)$, $GR_{apr\_2012, may\_2012}(P), GR_{apr\_2013, may\_2013}(P)$ are equal to 2 (0.5/0.25). While, the values of the support of the pattern P decrease through the pairs of the windows $[nov\_2011, dec\_2011]$, $[nov\_2012, dec\_2012]$ and $[nov\_2013, dec\_2013]$ and the values of growth-rate $GR_{dec\_2011, nov\_2011}(P), GR_{dec\_2012, nov\_2012}(P)$, $GR_{dec\_2013, nov\_2013}(P)$ are equal to 5. By supposing $minGR = 1.5$, the pattern P is considered emerging over the windows $[apr\_2011, may\_2011]$, $[nov\_2011, dec\_2011]$, $[apr\_2012, may\_2012]$, $[nov\_2012, dec\_2012]$, $[apr\_2013, may\_2013]$ and $[nov\_2013, dec\_2013]$. However, in the windows $[nov\_2011, dec\_2011]$, $[nov\_2012, dec\_2012]$ and $[apr\_2013, may\_2013]$ its variation of support is different from the variation detected in the windows $[apr\_2011, may\_2011]$, $[apr\_2012, may\_2012]$, $[apr\_2013, may\_2013]$ both in terms of quantity (5 against 2) and in terms of growth (decrease against increase). This means that we could build different periodic changes from P. Indeed, by supposing a function Θ which maps the values of growth-rate 2 and 5 to the nominal values *weak_change* and *strong_change*, the values of $minREP$ and δ equal 2 and 365 days respectively, we can generate two PCs having the same conjunction of weather parameters.

Definition 3. Spatio-temporal Periodic Change (SPC)

Let $T : \langle(\tau_{i_1}, \tau_{i_2}), \ldots, (\tau_{u_1}, \tau_{u_2})\rangle$ be a sequence of chronologically ordered pairs of time-windows, let $\Pi : \{PC_1 : \langle P, T_1, \psi\rangle, \ldots, PC_v : \langle P, T_v, \psi\rangle\}$ be a set of v periodic changes detected in v different geographic areal units, P is a spatio-temporal periodic change iff

1. $|\Pi| \geq minUNITS$
2. $\forall h \in \{i, \ldots, u\}, \forall k = 1, \ldots, v (\tau_{h_1}, \tau_{h_2}) \in T_k$
3. $\forall h \in \{i, \ldots, u-1\}(\tau_{h_1}, \tau_{h_2})$ and (τ_{k_1}, τ_{k_2}) are δ-separated, $k = h + 1$

Intuitively, a SPC represents a periodic variation (quantified by ψ) of the frequency of weather parameters conjunction P. Such a variation is observed in v different geographic areal units.

3   The Method

In this section we propose a method to mine SPCs from the measurements of the weather parameters $A_1, \ldots A_d$ recorded by sensors equally displaced over a geographic area on the sequence of time-points $\{t_1 \ldots t_n\}$. The method is structured in two steps performed consecutively (see Fig. 1). Initially, we build a gridded data space over the input geographic area in order to define the areal units as cells of equal size $\{c_{11}, \ldots, c_{\alpha, \beta}\}$. This means that the cells comprise the same number of sensors. The first step works on the values of the weather parameters of each cell c_{rs} and mines PCs in accordance with the Definition 2. The second step inputs the PCs detected on all the cells, it selects the PCs which are present in at least $minUNITS$ cells and then mines SPCs in accordance with the Definition 3. The details of these two steps are reported in the following.

3.1   Detection of Periodic Changes

To detect PCs, we adapt the algorithm proposed in [11] originally designed for data represented in relational logic, to the case of multi-dimensional time-series. In particular, it works on the succession $\langle(\tau_{1_1}, \tau_{1_2}), \ldots, (\tau_{h_1}, \tau_{h_2}), \ldots, (\tau_{z_1}, \tau_{z_2})\rangle$ of pairs of time-windows obtained from $\{t_1, \ldots, t_n\}$ (see Sect. 2). Each time-window τ_{u_v} (except the first and last one) is present in two consecutive pairs, so, given two pairs (τ_{h_1}, τ_{h_2}) and $(\tau_{(h+1)_1}, \tau_{(h+1)_2})$, we have that $\tau_{u_v} = \tau_{h_2} = \tau_{(h+1)_1}$. This is done to capture the changes of support of the patterns from τ_{h_1} to τ_{u_v} and from τ_{u_v} to $\tau_{(h+1)_2}$. The algorithm performs three main procedures.

1. Discovery of frequent patterns for each time-window. Frequent patterns are discovered from each time-window with the technique of evaluation-generation of candidate patterns used in [11], which exploits the monotonicity property of the support. Obviously, the decision of using that specific technique does not exclude the possibility of considering alternative solutions based on evaluation-generation of patterns, which do not imply modifications neither to our proposal nor to the set of frequent patterns resulting from the current procedure.

2. Extraction of the EPs from the frequent patterns discovered on τ_{h_1} against the frequent patterns discovered from τ_{h_2} in accordance with the Definition 1. To efficiently perform this operation, we can act on the support of the patterns. Indeed, we avoid the evaluation of a pattern $P2$, which is super-set of a pattern $P1$ ($P1 \subset P2$), if $P1$ is frequent in the time-window τ_{h_1} (τ_{h_2}) but it is not frequent in the time-window τ_{h_2} (τ_{h_1}). Instead, we cannot apply no optimization on the growth-rate because, unfortunately, the monotonicity property does not hold. In fact, given two frequent patterns $P1$ and $P2$ with $P1 \subset P2$, if $P1$ is not emerging, namely $GR_{\tau_{h_1},\tau_{h_2}}(P1) < minGR$ ($GR_{\tau_{h_2},\tau_{h_1}}(P1) < minGR$), then the pattern $P2$ may or may not be emerging, namely its growth-rate could exceed the threshold $minGR$.

   The final EPs are stored in a pattern base, which hence contains the frequent patterns that satisfy the constraint set by $minGR$ on at least one pair of time-windows. Each EP is associated with two lists, named as $TWlist$ and $GRlist$. $TWlist$ is used to store the pairs of time-windows in which the growth-rate of the pattern exceeds $minGR$, while $GRlist$ is used to store the corresponding values of growth-rate. The technical details can be found in the paper [11].

3. Detection of PCs from the EPs stored in the pattern base. To implement the function Θ_P (Definition 2) we resort to an equal-width discretization technique, which is able to return a set of ranges used here as nominal values Ψ. The discretization technique is applied to the set of values of the lists $GRlist$ of all the stored EPs. Thus, we can map a value of growth-rate to the range in which the value falls in. The choice of the equal-width discretization allows us to take the different magnitude orders into account and uniformly map the growth-rate values into different ranges, without making the distribution of the values unbalanced.

   The PCs are built with a procedure of generation-evaluation of candidates. In particular, we work on the EPs one at a time by generating as many candidates as the nominal values associated with the growth-rate of that EP. A PC is built incrementally by examining the pairs of time-windows of $TWlist$ in chronological order and joining those pairs that have the same nominal value ψ on the condition that they are δ-separated.

   In order to clarify how the detection of PCs works, we report an explanatory example of generation of PCs from one EP. Consider the time-points as years, $\Psi = \{\psi', \psi''\}$, $minREP = 3$, $\delta = 13$ and the lists $TWlist$ and $GRlist$ built as follows (the nominal value has the same position in $GRlist$ of the corresponding pair of time-windows in $TWlist$):

$TWlist : \langle([1970;1972],[1973;1975]),([1976;1978],[1979;1981]),([1982;1984],[1985,1987]),$
$([1988;1990],[1991;1993]),([1994;1996],[1997;1999]),([2010;2012],[2013;2015])\rangle$
$GRlist : \psi',\psi',\psi'',\psi',\psi'',\psi'\rangle$

By scanning the list $TWlist$, we can initialize the sequence T of a candidate PC' by using the pairs ([1970;1972], [1973;1975]) and ([1976;1978], [1979;1981]) since they are δ-separated (1979–1973 < δ) and they have the same nominal value ψ'. The pair ([1982;1984], [1985;1987]) instead refers to a different nominal value (ψ'') and therefore it cannot be inserted into T of PC'. We use it to initialize the sequence T of a new candidate PC'', which thus will include the time-windows referred to ψ''. Subsequently, the pair ([1988;1990], [1991;1993]) is inserted into T of PC' since its distance from the latest pair is less than δ (1991–1979 < δ). Then, T of PC'' is updated with ([1994;1996], [1997;1999]) since 1997–1985 is less than δ, while the pair ([2010;2012], [2013;2015]) cannot be inserted into T because the distance between 2013 and 1997 is greater than δ. Thus, we use the pair ([2010;2012], [2013;2015]) to initialize the sequence T of a new candidate PC'''. The sequence T of PC' cannot be further updated, but, since its size exceeds $minREP$, we consider the candidate PC' as valid periodic change. Finally, the candidate PC'' cannot be considered as valid since its size is less than $minREP$. The candidate PC''' is not even considered since its sequence T has less than $minREP$ elements.

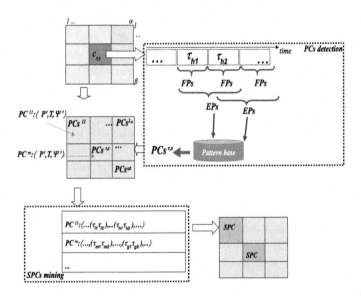

Fig. 1. The block-diagram of the two-step method for mining spatio-temporal patterns of periodic changes.

3.2 Mining Spatio-Temporal Periodic Changes

As result of the first step, we have a set of PCs for each cell. A preliminary operation we perform is the removal of redundant PCs. Indeed, the invalidity of the property of monotonicity of the growth-rate and the procedure of detection of PCs do not allow us to exclude the presence of redundancies, that is, PCs whose information is expressed also by other PCs. For instance, given two PCs, PC': $\langle P', T', \psi \rangle$ and PC": $\langle P'', T'', \psi \rangle$, P' is redundant if (i) the conjunction of weather parameters of P'' includes the conjunction of weather parameters of P' ($P' \subset P''$); (ii) the pairs of time-windows of PC" comprise those of PC' ($T' \subset T''$); (iii) they have the same nominal value ψ.

After having removed the redundant PCs, to find SPCs we should act on the sequences T. Different alternatives can be considered, which we discuss briefly in the following. Using a grouping/clustering algorithm could turn out to be inapplicable because the lengths of T can be different. This is also the reason for which we cannot adopt algorithms for the generation of frequent itemsets. The distance-based techniques, for instance those implementing the dynamic-time-warping distance, could be ineffective because, although able to handle sequences of different lengths, they return groups of sequences with similar/close time-windows, whilst we are interested in obtaining sequences with identical time-windows. Our proposal is investigating this problem with a sequence mining approach, which naturally handles sequences of different lengths and takes the chronological order of the time-windows into account [14]. Here, the input data of the sequence mining problem is the set of the sequences T of one PC in common to several cells, for instance $\{PC^{11}, \ldots, PC^{rs}\}$ in Fig. 1. So, we take the set of the sequences T associated with a specific emerging pattern P' having a specific nominal value ψ'. The output is the complete set of SPCs in form of sequential patterns whose elements are pairs of time-windows. By considering that there are different PCs, the algorithm of sequence mining is applied to one collection of sets of sequences, whose cardinality is equal to the total number of PCs. Not all the PCs are used for the sequence mining algorithm but only those found in at least $minUNITS$ cells.

Here, we could experience the problem of redundant patterns, so we decide to use an algorithm able to mine *closed* sequential patterns. A sequential pattern S' is closed if there exists no sequential pattern S'' such that $S' \subset S''$ and S'' occurs in the same sequences of S'. The use of closed sequential patterns allows us to additionally maximize both the number of cells in which the change occurs and number of repetitions of the change in each cell. We exploit the algorithm CloSpan [18], which implements a candidate maintenance-and-test approach. It first generates a set of closed sequence candidates, which is stored in a hash-based tree structure and then performs a post-pruning operation on that set. The post-pruning operation exploits search space techniques. Obviously, the decision of using the algorithm CloSpan does not exclude the possibility of considering alternative solutions. Indeed, other algorithms of closed sequential patterns mining do not imply modifications to the method, considering that our purpose here is the generation of the minimal set of frequent sequences of pairs of the windows for each periodic change.

Finally, not all the closed sequential patterns are considered but only those that meet two conditions: (i) the pairs of time-windows are δ-separated and (ii) the grid cells associated to the patterns are adjacent. These cells denote together the spatial region in which a periodic change occurs.

4   Experiments

We applied the proposed method to real-world climate data generated from the NCEP/NCAR Reanalysis project and available on the data bank NOAA [15]. The climate data were recorded every day from January 1997 to December 1999 by 697 sensors uniformly distributed over a grid of 41×17 points (41 sensors by longitude, 17 sensors by latitude). So, totally we have 1094 daily measurements (1094 time-points). The distribution of the sensors delimits a specific geographic area localized between Atlantic Ocean and Indian Ocean and covers almost $36,000,000\,km^2$. The weather parameters are "Air temperature", "Pressure", "Relative humidity", "Eastward Wind", "Northward Wind" and "Precipitable Water".

Experimental Setup. We pre-processed the time-series by using an equal-frequency discretization technique, which guarantees a uniform distribution on the five (discretized) ranges of the same parameter when generating patterns with the ranges of different parameters. In particular, for each parameter, we considered 5 ranges. To implement the function Θ_P, we applied an equal-width discretization technique to the values of the growth-rate experimentally obtained, which fall in the interval $[1.2, 5]$. The number of the ranges generated is 6, namely $\{[1.5, 2), [2, 2.5), [2.5, 3), [3, 3.5), [3.5, 4), [4, 4.5)\}$, to which we manually assign the nominal values $very\_weak\_change, weak\_change, middle\_weak\_change, \quad middle\_strong\_change, strong\_change, very\_strong\_change$. So, the function Θ_P maps values from the interval $[1.2, 5]$ to the set $\{very\_weak\_change, weak\_change, middle\_weak\_change, \quad middle\_strong\_change, strong\_change, very\_strong\_change\}$.

We built three different configurations of the grid from the geographic area. In each configuration, the grid cells cover the same number of sensors and therefore have the same size. Specifically, the distribution of the sensors in each cell is 10×8, 5×8, 8×4, respectively, so the three configurations have 8 cells, 16 cells, 20 cells. Experiments are performed by tuning $minGR, \delta$ and $minREP$. The value of minimum support for the step of PCs detection is fixed to 0.1, while the value of minimum support for the step of SPCs mining is fixed to 0.5 in order to find patterns which cover at least the half of the minimum number of cells fixed by $minUNITS$. The value of $minUNITS$ equals the half of the total number of cells for each grid configuration, that is, 4, 8, 10 respectively. The value of the width w of the windows is 30 (days).

Results. We collected three kinds of quantitative results. Specifically, Table 1 illustrates the values of PCs averaged by the number of cells and the total number of SPCs. Table 2 reports the evaluation of the SPCs in form of average portion of cells in which the final SPCs occur. More precisely, the evaluation considers the number of cells covered by the SPCs divided by the minimum number of requested cells ($minUNITS$) and has values in $[0;1]$, where 1 refers the best coverage and indicates that the SPCs cover all the cells provided by $minUNITS$. In the following, we discuss the influence of the input thresholds $minGR, \delta$ and $minREP$ on these results.

Discussion. In the boxes (a), (b) and (c) of Table 1, we report the results obtained with the three grid configurations. We see that the smaller the area of the cells the lower the number of PCs and SPCs, meaning that the method is able to capture a quite expected behavior, that is, the spatial regions with greater extent show there higher variability of the weather conditions compared with the smaller regions. As to the influence of $minGR$, we observe that there not are PCs and SPCs when it is higher than 6. This indicates that there is no conjunction of weather parameters whose frequency increases or decreases by an order of magnitude higher of 6.

By increasing only the threshold δ, we have greater sets of PCs. Indeed, at higher values of δ the method detects both the changes which are replicated more frequently (that is, at $\delta = 60$ days) and the changes which are replicated less frequently, that is, with distant repetitions ($\delta = 365$). Consequently, the sets of the PCs (which are the input of the step of SPCs mining) are greater and this implies the discovery of greater sets of SPCs.

By increasing only the threshold $minREP$, we obtain smaller sets of PCs. In fact, when setting higher values of $minREP$, we require climate changes with a relatively high number of repetitions, which is a requirement that only the PCs with longer sequences of T can satisfy. Consequently, the number of PCs that feeds the second step (SPCs mining) is lower and the set of SPCs is smaller but it is composed by the longer SPCs since generated with longer PCs. This is evident whether comparing the tables in the box (a) against those in the boxes (b) and (c). A concrete example is when $minREP$ is 5 ($w = 90$ days). In that case, we search changes repeated at a distance of even 5 semesters, that is, almost the whole dataset (6 semester long).

Table 2 reports a quantitative evaluation of the SPCs. We see that the better coverage is almost three-quarters of the requested cells and it is reached at the lowest values of $minGR$ and $minREP$ and highest value of δ. By considering only $minGR$, we observe that the better result is obtained at $minGR = 1.5$, which corresponds to SPCs with "very weak changes". Instead, when $minGR > 4$, we have SPCs with "very strong changes" but replicated in a smaller set of cells. By considering only δ, we note that there is a discrete coverage of the cells at relatively low values δ. This can be explained by the lower number of SPCs. Finally, by increasing only the threshold $minREP$, the coverage decreases because of the combined effect of the number of the SPCs and their length. This is not surprising because weather changes with less repetitions occur in larger spatial regions, while those with more repetitions are present in smaller regions.

Table 1. Results obtained by tuning a parameter at time on three grid configurations, that is, 8 cells (a), 16 cells (b) and 20 cells (c). When tuning $minGR$, δ is 365 and $minREP$ is 3. When tuning δ, $minGR$ is 2 and $minREP$ is 3. When tuning $minREP$, $minGR$ is 2 and δ is 365. Each slot of the tables reports the average values of PCs and the total number of SPCs. The average values of PCs are computed on the number of the cells.

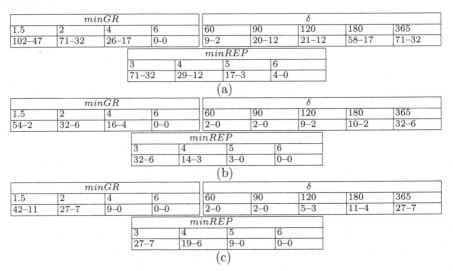

minGR				δ				
1.5	2	4	6	60	90	120	180	365
102–47	71–32	26–17	0-0	9–2	20–12	21–12	58–17	71–32

minREP			
3	4	5	6
71–32	29–12	17–3	4–0

(a)

minGR				δ				
1.5	2	4	6	60	90	120	180	365
54–2	32–6	16–4	0-0	2–0	2–0	9–2	10–2	32–6

minREP			
3	4	5	6
32–6	14–3	3–0	0–0

(b)

minGR				δ				
1.5	2	4	6	60	90	120	180	365
42–11	27–7	9–0	0-0	2–0	2–0	5–3	11–4	27–7

minREP			
3	4	5	6
27–7	19–6	9–0	0–0

(c)

Table 2. A quantitative evaluation of the SPCs in terms of average portion of distinct cells covered by the final SPCs.

minGR				δ				
1.2	2	4	6	60	90	120	180	365
0.72	0.71	0.52	–	0.55	0.53	0.59	0.66	0.71

minREP			
3	4	5	6
0.71	0.56	0.51	–

Interpretation of the Spatio-Temporal Patterns. Here we present some examples of SPCs mined from the real-word climate data and report the pairs of windows over which they are repeated and the modelled change. The grid cells are graphically drawn on the geographic map for ease of the interpretation.

For instance, the following SPC has been mined with $minGR = 2$, $\delta = 365$, $minREP = 3$

$$SPC_1 : [P : air\_temperature = [301.5; 307.2], pressure = [96, 99; 100],$$

$$relative\_humidity = [82.75; 89.75], precipitable\_water = [0.46; 10.89];$$

$$T : \langle([june\_1997, july\_1997], [may\_1998, june\_1998]), ([may\_1999, june\_1999])\rangle;$$

$$\Psi = middle\_weak\_change]$$

SPC_1 represents a change of frequency denoted as $middle\_weak\_change$, which corresponds to the range [2.5;3]. This variation recurs three times, specifically over the pairs of windows $\langle([june\_1997, july\_1997], [may\_1998, june\_$ $1998]), ([may\_1999, june\_1999])\rangle$ and it is replicated on the five cells drawn Fig. 2a. Intuitively, we see that such a change recurs with a periodicity of at most 12 months and covers the land of the geographic area under examination.

Another SPC the method discovered with $minGR = 2$, $\delta = 365$, $minREP = 3$ is the following

$$SPC_2 : [P : air\_temperature = [301.5; 307.2], pressure = [96, 99; 100],$$
$$relative\_humidity = [82.75; 89.75];$$
$$T : \langle([june\_1997, july\_1997], [may\_1998, june\_1998]), ([may\_1999, june\_1999])\rangle;$$
$$\Psi = middle\_strong\_change]$$

It exhibits a frequency variation included in the range $[3; 3.5]$ (*middle_strong_change*) over the pairs of windows $T : \langle([june\_1997, july\_1997], [may\_1998,$ $june\_1998]), ([may\_1999, june\_1999])\rangle$ and on the five cells drawn in dashed in Fig. 2b. We see that SPC_2 has a conjunction of weather parameters, which is a subset of SPC_1, additionally, it appears in the same sequence and same spatial region of SPC_1, but it denotes a stronger change. This means that the support of SPC_2 is higher than the support of SPC_1 either in the windows $june\_1997$, may_1998, may_1999 or in the windows $july\_1997$, $june\_1998$, $june\_1999$. This is explained with the monotonicity property of the support.

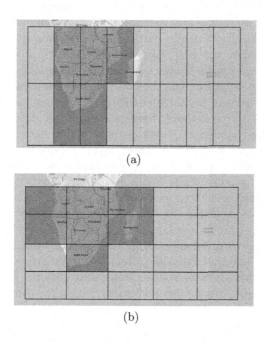

(a)

(b)

Fig. 2. Visualization of the grid obtained by collecting 5×8 (a) and 8×4 (b) sensors per cell. The spatio-temporal periodic changes SPC_1 and SPC_2 occur in the dashed cells of the grid (a). The spatio-temporal periodic changes SPC_2 occurs also in the dashed cells of the grid (b).

The same periodic change represented by SPC_2 has been mined with $minGR = 2$, $\delta = 365$, $minREP = 3$ by using the grid configuration with 20 cells (Fig. 2b). In this case, it has the same pairs of windows T that had within the grid at 16 cells and clearly covers a different subset of cells. We see that the spatial region of SPC_2 in the first grid (Fig. 2a) greatly overlaps the spatial region of SPC_2 in the second grid (Fig. 2b).

5   Related Work

The analysis of climate data has always attracted interest by different disciplines and the study of the dynamics is considered particularly relevant for the effects on the Earth. Günnemann et al. [6] work on the hypothesis that the changes can regard subspaces of the descriptive attributes. Then, they describe a clustering technique based on the similarity which tracks the changes of subspaces in time-variable climate data and associates a type of climate behaviour with each cluster. Kleynhans et al. [8] propose a method to detect and evaluate land cover change by examining at each point in time for a specific pixel neighborhood the spatial covariance of a hyper-temporal time series. McGuire et al. [13] introduce the problem of mining moving dynamic regions. Their solution is based on spatial auto-correlation and finds dynamic spatial regions across time periods and dynamic time periods over space. Finally, moving dynamic regions are identified by determining the spatio-temporal connectivity, extent, and trajectory for groups of locally dynamic spatial locations whose position has shifted from one time period to the next. Lian and McGuire [9] propose an algorithm to detect high change regions based on quadtree-based index and classify heterogeneous and homogeneous change. Finally, spatio-temporal changes are analyzed at long time scales to find high change persistent regions and high change dynamic regions. In [1], the authors investigate a problem of change analysis with a descriptive method aiming at summarizing evolving data streams in spatial domains. They propose a clustering-based technique to detect groups of georeferenced data which vary according to a similar trend, which is determined over time-windows.

The periodicity has been often seen as a disturbance effect to be removed from the climate data because makes the applicability of the classical methods unfeasible. Tan et al. [16] present a comprehensive study based on classical pattern discovery algorithms to find spatio-temporal patterns from spatial zones over time. Preliminarily, seasonal variation is removed from data with data transformation techniques, like discrete Fourier transform. Patterns denote regularities within individual zones, among different zones, within the same time-interval or along a series of time-intervals. The study presented in [10] focused on the periodic variation of phenotype data and applied the solution to seasonal diseases. However, as our knowledge, very few attempts have been done to investigate the periodicity of the change over space and no attempt focused on the use of patterns. Boriah et al. [2] proposed a recursive merging algorithm that exploited the seasonality to distinguish between locations that experienced a land cover change and locations that did not. However, it does provide no information on the change and on the spatial and temporal components associated to it. In [12] the authors investigated the effect of the periodicity in form of temporal auto-correlation for regression problems on time-stamped networks. Spatio-temporal patterns are the main subject of study in trajectory mining. In [7] the authors propose unifying incremental approaches to automatically extract different kinds of spatio-temporal patterns by applying frequent closed item-set mining techniques.

6   Conclusions

The research presented in this paper has two main contributions. First, we extend a previous method, in order to identify different occurrences of the same periodic changing behavior. Second, we explore the possibility to identify periodic changing behaviors in Climatology, which is typically characterized by temporal and spatial component. We have introduced the notion of spatial-temporal pattern of periodic changes to denote the spatial extent of variations repeated on the temporal axis. The proposed method relies on the frequent pattern mining framework, which enables us to *(i)* capture the changes in terms of variations of the frequency, *(ii)* estimate the regularity over time of these changes, and *(iii)* identify contiguous areal units in which the change can be tracked. The application to a real dataset highlights the viability and usefulness of the proposed method to a real-world problem. We performed experiments to test the sensibility of the method with respect to input thresholds. We plan to explore different future directions: (i) automatic determination of the input parameters, (ii) qualitative evaluation the discovered patterns against ground-truth on weather changes (iii) study of the usefulness of the patterns for predictive problems.

Acknowledgements. The authors would like to acknowledge the support of the European Commission through the project MAESTRA - Learning from Massive, Incompletely annotated, and Structured Data (Grant number ICT-2013-612944).

References

1. Appice, A., Ciampi, A., Malerba, D.: Summarizing numeric spatial data streams by trend cluster discovery. Data Min. Knowl. Discov. **29**(1), 84–136 (2015)
2. Boriah, S., Kumar, V., Steinbach, M., Potter, C., Klooster, S.: Land cover change detection: a case study. In: Proceedings of the 14th ACM SIGKDD International Conference on Knowledge Discovery and Data Mining, KDD 2008, pp. 857–865. ACM, New York (2008)
3. Chandola, V., Banerjee, A., Kumar, V.: Anomaly detection for discrete sequences: a survey. IEEE Trans. Knowl. Data Eng. **24**(5), 823–839 (2012)
4. Dong, G., Li, J.: Efficient mining of emerging patterns: discovering trends and differences. In: Proceedings of the Fifth ACM SIGKDD International Conference on Knowledge Discovery and Data Mining, pp. 43–52 (1999)

5. Faghmous, J.H., Kumar, V.: Spatio-temporal data mining for climate data: advances, challenges, and opportunities. In: Chu, W.W. (ed.) Data Mining and Knowledge Discovery for Big Data. Studies in Big Data, vol. 1, pp. 83–116. Springer, Heidelberg (2014). doi:10.1007/978-3-642-40837-3_3

6. Günnemann, S., Kremer, H., Laufkötter, C., Seidl, T.: Tracing evolving subspace clusters in temporal climate data. Data Min. Knowl. Discov. **24**(2), 387–410 (2012)

7. Hai, P.N., Poncelet, P., Teisseire, M.: GET_MOVE: an efficient and unifying spatio-temporal pattern mining algorithm for moving objects. In: Hollmén, J., Klawonn, F., Tucker, A. (eds.) IDA 2012. LNCS, vol. 7619, pp. 276–288. Springer, Heidelberg (2012). doi:10.1007/978-3-642-34156-4_26

8. Kleynhans, W., Salmon, B.P., Wessels, K.J.: A novel spatio-temporal change detection approach using hyper-temporal satellite data. In: 2014 IEEE Geoscience and Remote Sensing Symposium, IGARSS 2014, Quebec City, QC, Canada, 13–18 July 2014, pp. 4208–4211. IEEE (2014)

9. Lian, J., McGuire, M.P.: Mining persistent and dynamic spatio-temporal change in global climate data. In: Latifi, S. (ed.) Information Technology: New Generations. AISC, vol. 448, pp. 881–891. Springer, Cham (2016). doi:10.1007/978-3-319-32467-8_76

10. Loglisci, C., Balech, B., Malerba, D.: Discovering variability patterns for change detection in complex phenotype data. In: Esposito, F., Pivert, O., Hacid, M.-S., Raś, Z.W., Ferilli, S. (eds.) ISMIS 2015. LNCS (LNAI), vol. 9384, pp. 9–18. Springer, Cham (2015). doi:10.1007/978-3-319-25252-0_2

11. Loglisci, C., Malerba, D.: Mining periodic changes in complex dynamic data through relational pattern discovery. In: Ceci, M., Loglisci, C., Manco, G., Masciari, E., Ras, Z.W. (eds.) NFMCP 2015. LNCS (LNAI), vol. 9607, pp. 76–90. Springer, Cham (2016). doi:10.1007/978-3-319-39315-5_6

12. Loglisci, C., Malerba, D.: Leveraging temporal autocorrelation of historical data for improving accuracy in network regression. Stat. Anal. Data Min. **10**(1), 40–53 (2017)

13. McGuire, M.P., Janeja, V.P., Gangopadhyay, A.: Mining trajectories of moving dynamic spatio-temporal regions in sensor datasets. Data Min. Knowl. Discov. **28**(4), 961–1003 (2014)

14. Mooney, C.H., Roddick, J.F.: Sequential pattern mining - approaches and algorithms. ACM Comput. Surv. **45**(2), 19:1–19:39 (2013)

15. Simons, R.A.: ERDDAP - the environmental research division's data access program. NOAA/NMFS/SWFSC/ERD, Pacific Grove (2011). http://coastwatch.pfeg.noaa.gov/erddap

16. Tan, P., Steinbach, M., Kumar, V., Potter, C., Klooster, S., Torregrosa, A.: Finding spatio-temporal patterns in earth science data. In: Proceedings of KDD Workshop on Temporal Data Mining (2001)

17. Wilby, R.L., Wigley, T.M.L.: Downscaling general circulation model output: a review of methods and limitations. Prog. Phys. Geogr. **21**(4), 530–548 (1997)

18. Yan, X., Han, J., Afshar, R.: CloSpan: mining closed sequential patterns in large databases. In: Proceedings of the Third SIAM International Conference on Data Mining, CA, USA, 1–3 May 2003, pp. 166–177 (2003)

Mining Keystroke Timing Pattern for User Authentication

Saket Maheshwary$^{(\boxtimes)}$ and Vikram Pudi

Center for Data Engineering and Kohli Center on Intelligent Systems,
International Institute of Information Technology-Hyderabad, Hyderabad, India
saket.maheshwary@research.iiit.ac.in, vikram@iiit.ac.in

Abstract. In this paper we investigate the problem of user authentication based on keystroke timing pattern. We propose a simple, robust and non parameterized nearest neighbor regression based feature ranking algorithm for anomaly detection. Our approach successfully handle drawbacks like outlier detection, scale variation and prevents overfitting. Apart from using existing keystroke timing features from the dataset like dwell time and flight time, other features namely bigram time and inversion ratio time are engineered as well. The efficiency and effectiveness of our method is demonstrated through extensive comparisons with other state-of-the-art techniques using CMU keystroke dynamics benchmark dataset and has shown great results in terms of average equal error rate (EER) than other proposed techniques. We achieved an average equal error rate of **0.051** for the user authentication task.

Keywords: Anomaly detection · Feature ranking · Nearest neighbor · Regression · Prediction · Security

1 Introduction

In this era where everyone wants secure, faster, reliable and easy to use means of communication, there are many instances where user information such as personal details and passwords get compromised thus posing a threat to system security. In order to tackle the challenges posed on the system security biometrics [8] prove to be a vital asset. Biometric systems are divided into two classes namely physiology based ones and the ones based on behavior. Physiology based approach allows authentication via use of retina, voice and fingerprint touch. In contrast, behavior based approach includes keystroke dynamics on keyboard or touch screens and mouse click patterns.

In this paper we propose a learning model to deal with *keystroke dynamics* – a behavior based unique timing patterns in an individuals typing rhythm which is used as a protective measure. These rhythms and timing patterns of tapping are idiosyncratic [1] the same way as handwriting or signatures are, due to their similar governing neurophysiological mechanisms. Back in the 19th century, telegraph operators could recognize each other based on ones specific tapping

© Springer International Publishing AG 2017
A. Appice et al. (Eds.): NFMCP 2016, LNAI 10312, pp. 213–227, 2017.
DOI: 10.1007/978-3-319-61461-8_14

style [15]. Based on the analysis of the keystroke timing patterns, it is possible to differentiate between actual user and an intruder. By keystroke dynamics we refer to any feature related to the keys that a user presses such as key down time, key up time, flight time etc. In this paper, we concentrate on classifying users based on static text such as user password. The mechanism of keystroke dynamics can be integrated easily into existing computer systems as it does not require any additional hardware like sensors thus making it a cost effective and user friendly technique for authenticating users with high accuracy. It is appropriate to use keystroke dynamics for user authentication as studies [21,22] have shown that users have unique typing patterns and style. Moreover [21,22] has proven some interesting results in their research work as well. First, [21,22] proved is that the users present significantly dissimilar typing patterns. Second they have shown details about the relationship between users occurrence of sequence of events and their typing style and ability. Then [21,22] explained sequence of key up and key down events on the actual set of keys. Then [21,22] have also shown that there is no correlation between users typing skills and the sequence of events. Hence all these factors make it difficult for intruders to match with the actual users typing patterns. Keystroke dynamics is concerned with users timing details of typing data and hence various features could be generated from these timing patterns. In this paper we are using timing features only on static text.

The rest of the paper is organized as follows. In Sect. 2 we discuss related work and our contribution. In Sect. 3 we discuss the details of how manual features are engineered from the dataset and in Sect. 4 we discuss the concept of optimal fitting line. In Sect. 5 we present our proposed algorithm for feature ranking which is divided into two sub sections where first subsection discusses proposed approach on how feature ranking is done using nearest neighbor regression and second subsection discusses the neural network we used on the ranked or weighted feature space for anomaly detection. In Sect. 6 we experimentally evaluate our algorithm and show the results. Finally, we conclude our study and identify future work in Sect. 7.

2 Related Work

Classifying users based on keystroke timing patterns has been in limelight when [6] first investigated whether users could be distinguished by the way they type on keyboard. Researchers have been studying the user typing patterns and behavior for identification. Then [7] investigated the possibility of using keystroke timings as to whether typists could be identified by analyzing keystroke times as they type long passages of text. Later [17] extracted keystroke features using the mean and variance of digraphs and trigraphs. A detailed survey [18] on the keystroke dynamics literature using the Euclidean distance metric with Bayesian like classifiers. Initially [3] and later [9] proposed to use the relative order of duration times for different n-graphs to extract keystroke features that was found to be more robust to the intra-class variations than absolute timing. Also [9] published great results for text-free keystroke dynamics identification where they merge relative and absolute timing information on features. Then [23] proposed a new distance metric by combining Mahalanobis and Manhattan distance metrics. Many machine learning techniques have been proposed as well for keystroke dynamics as an authentication system. Keystroke dynamics can be applied with variety of machine learning algorithms like Decision Trees, Support Vector Machines, Neural Networks, Nearest Neighbor Algorithms [5] and Ensemble Algorithms [19] among others.

One problem faced by researchers working on these type of problems is that majority of the researchers are preparing their own dataset by collecting data via different techniques and the performance criteria is not uniform as well hence comparison on similar grounds among the proposed algorithms becomes a difficult task. To address this issue, keystroke dynamics benchmark dataset is publicly provided with performance values of popular keystroke dynamics algorithms [12] to provide a standard universal experimental platform. They collected and published a keystroke dynamics benchmark dataset containing 51 subjects with 400 keystroke timing patterns collected for each subject. Besides this they also evaluated fourteen available keystroke dynamics algorithms on this dataset, including Neural Networks, KNNs, Outlier Elimination, SVMs etc. Various distance metrics including Euclidean distance, Manhattan distance and Mahalanobis distance were used. This keystroke timing pattern dataset along with the evaluation criteria and performance values stated provides a benchmark to compare the progresses of new proposed keystroke timing pattern algorithms on same grounds.

2.1   Our Contribution

The performance study of the fourteen existing keystroke dynamics algorithms implemented by [12] indicated that the top performers are the classifiers using scaled Manhattan distance and the nearest neighbor classifier. In this paper we present a new nearest neighbor regression based feature ranking algorithm for anomaly detection that assigns weight to the feature vector.

Nowadays neural network based models are frequently used in the field of computer vision, speech signal processing, text representation have now been adopted in the fields of security as well. These neural network based techniques have multiple advantages over previous approaches both in task specific performance and scalability. Motivated by the superior results obtained by the neural network models we decided to use it as a classifier for anomaly detection. We used a simple 3 layer neural network classifier for anomaly detection by giving the weighted feature space generated by our model as input to neural network. Our proposed approach has the following desirable features:

- **Parameterless:** We first design our nearest neighbor based regression algorithm and then show how the parameter can be automatically set, thereby resulting in a parameterless algorithm. This removes the burden from the user of having to set parameter values – a process that typically involves repeated trial-and-error for every application domain and dataset.

- **Accurate:** Our experimental study in Sect. 6 shows that our algorithm provides more accurate estimates than its competitors. We compare our approach with 14 other algorithm using the same evaluation criteria for objective comparison.
- **Robust/Outlier Resilient:** Another problem with the statistical approaches is outlier sensitivity. Outliers (extreme cases) can seriously bias the results by pulling or pushing the regression curve in a particular direction, leading to biased regression coefficients. Often, excluding just a single extreme case can yield a completely different set. The output of our algorithm for a particular input record is dependent only on its nearest neighbors hence insensitive to far-away outliers.
- **Simple:** The design of our algorithm is simple, as it is based on the nearest neighbor regression. This makes it easy to implement, maintain, embed and modify as the situation demands.

Apart from our proposed algorithm we have engineered two new features namely *Bigram time* and *Inversion Ratio time* as discussed in Sect. 3.

3   Feature Engineering

What are good timing features that classify a user correctly? This is still an open research problem. Though keystroke up, keystroke down and latency timing are the commonly used features, in this paper we have generated two new features from the given dataset besides the existing features. The dataset [12] provides three types of timing information namely the hold time, key down-key down time and key up-key down time. Besides these three existing features, two new features namely Bigram time and Inversion ratio time are engineered. Following are the details of five categories of timing features which is used to generate 51 features using keystroke timing dataset [12]. Figure 1 illustrates various timing features where up arrow indicates key press and down arrow indicates key release.

- **Hold Time** also known as dwell time, is the duration of time for which the key is held down i.e. the amount of time between pressing and releasing a single key. In Fig. 1, H_i represents the hold time.
- **Down-Down Time** key down key down time is the time from when key1 was pressed to when key2 was pressed. In Fig. 1, the times DD_i depicts the down time.
- **Up-Down Time** key up key down time is the time from when key1 was released to when key2 was pressed. This time can be negative as well. In Fig. 1, the times UD_i depicts the up down time.
- **Bigram Time** is the combined time taken by two adjacent keystrokes i.e. the time from pressing down of key1 to releasing to key2.
- **Inversion Ratio Time** it is the timing ratio of hold time of key1 and key2 where key1 and key2 are the two continuous keystrokes. In Fig. 1, H_{i+1}/H_i is the inversion ratio time.

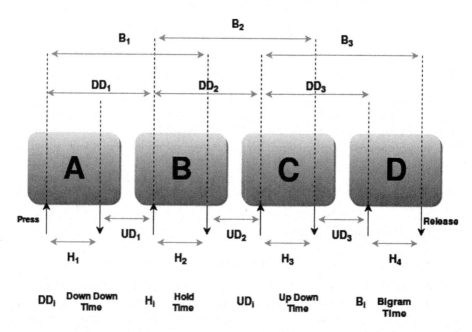

Fig. 1. Illustration of generated keystroke timing features where A, B, C, D are the keys

4 Optimal Fitting Line

Regression algorithms are used for predicting (time series data, forecasting), testing hypothesis, investigating relationship between variables etc. Here in this section we discuss how the optimal fitting line attempts to predict the relationship between one variable from one or more other variables by fitting a linear equation to observed data.

In this paper we assume that to construct the line of best fit, with increase or decrease in each independent variable value the dependent variable changes smoothly. Thus this helps us in achieving almost linear relationship between dependent and independent variables thus allowing us to optimally fit a line onto the points in a small neighborhood. The line which minimizes the mean squared error is referred to as optimal fitting line. A low value of error indicates that the line is optimally fitted to the neighborhood and has captured the linearity of the locality. Let the k points have values $\{(x_1, y_1), ..., (x_k, y_k)\}$ in dimension x and y and let the variable to be predicted be y. Let the equation of line be of the form $y = ax + b$. Hence, dependent variable will take the value $ax_i + b$ corresponding to tuple i. Let the error in prediction for tuple i be denoted as e_i and is equal to $|y - ax_i - b|$. Hence the local mean squared error (LME) is denoted as,

$$LME(a, b) = \frac{\sum_{i=1}^{k} e_i}{k} = \frac{\sum_{i=1}^{k}(y - ax_i - b)^2}{k} \tag{1}$$

By minimizing LME where a and b are the parameters denoted by,

$$a = \frac{\sum_{j=1}^{k} y_j \sum_{i=1}^{k} x_i - k \sum_{i=1}^{k} x_i y_i}{\sum_{j=1}^{k} x_j \sum_{i=1}^{k} x_i - k \sum_{i=1}^{k} x_i^2} \tag{2}$$

$$b = \frac{\sum_{i=1}^{k} y_i - a \sum_{i=1}^{k} x_i}{k} \tag{3}$$

Thus, we get the equation of the optimal fitting line. Now after constructing the line of best fit, we are able to predict the dependent values for test tuple. Then we compare the actual and the predicted values of dependent variable to calculate least mean error for the given test tuple. Now based on the mean error, we are assigning weights to our feature vector in inverse proportion which is discussed in Sect. 5.

5   Our Proposed Approach

5.1   Support for Categorical Attributes

In this section we discuss how our proposed approach deals with categorical data. The keystroke timing dataset that we used for evaluating our approach has categorical attributes. One of the serious limitations of existing regression algorithms is their support only for numeric attributes. So in order to tackle this problem we are using a similarity measure which helps us to quantify the relation between two classes using some real valued function. The section below explains the similarity function that we have used in this paper. Most of real life datasets have mixed attributes (set of both numeric and categorical type attributes) and hence to overcome this situation we are using *cosine similarity* measure. The dot product for two vectors $\boldsymbol{A} = (a_1, a_2, ...)$ and $\boldsymbol{B} = (b_1, b_2, ...)$ where a_n and b_n are the components of the vector and n is the dimension of the vector space. Hence the dot product between A and B is formulated as $\boldsymbol{A} \cdot \boldsymbol{B} = a_1 b_1 + a_2 b_2 + ... + a_n b_n$.

The cosine similarity between two vectors is a measure that calculates the cosine of the angle between them. This metric is a measurement of orientation and not magnitude, it can be seen as a comparison between timing vectors on a normalized space because we are not taking into consideration only the magnitude of each timing vector, but the angle between them. What we have to do to build the cosine similarity equation is to solve the equation of the dot product for the $\cos \theta$. Thus, the similarity values obtained using cosine similarity gives us a clear estimate of how similar the categorical values are with respect to the class labels.

$$\cos \theta = \frac{\boldsymbol{A} \cdot \boldsymbol{B}}{||\boldsymbol{A}|| \, ||\boldsymbol{B}||} \tag{4}$$

5.2   Proposed Algorithm

In this section we discuss our proposed nearest neighbor regression algorithm in detail. Our algorithm successfully eliminates nearest neighbor algorithm problems like choice of number of neighbors k by choosing the optimal k value corresponding to minimum error thus making our algorithm to be non parametric in nature. Our algorithm uses a unique weighing criteria (Algorithm 2) to assign weights to the feature vector hence enabling us to determine the relative importance of dimensions. The notation used for the algorithm is as follows: The training data has d dimensions with feature variables $(A_1, A_2, ..., A_d)$ and the value of the feature variable for the jth feature variable A_j corresponding to the ith tuple can be accessed as $data[i][j]$. The value of the dependent variable of the training tuple corresponding to id value i can be accessed as $y[i]$. The value of the dependent variable is calculated using the cosine similarity and k represents the number of nearest neighbors. For a given test tuple T the value of its k nearest neighbors is determined using an iterative procedure (line 4 of Algorithm 1) hence making our algorithm to be non parametric in nature. The range for value k is from *low* to *high* where *low* is set to value 5 (sufficiently small value) an *high* is set to *size of training data data/2* (sufficiently large value). Now we describe our algorithm using the pseudo code below shown in Algorithms 1 and 2.

We iterate for k in range *5* to *size of training data set/2* and calculate the k nearest neighbors for test data. The k evaluated neighbors are stored in list *ClosestNeighbors* (line 6 of Algorithm 1). Now Algorithm 2 constructs an optimal fitting line $Line_i$ for each dimension of our feature vector (the dataset used by us has 51 features) by fitting a linear equation to observed *ClosestNeighbors* list, in the plane of feature variable and the dependent variable. The regression line is constructed as discussed in Sect. 4. Using the parameters from the equation of the line a and b (Eqs. 2 and 3) we predict the dependent value of test data (line 4 of Algorithm 2). Based on the predicted and actual values of the dependent variable squared error E_i is calculated (line 5 of Algorithm 2).

Algorithm 1

1: **procedure** KNN BASED DIMENSIONAL REGRESSION
2:   $MinimumError \leftarrow \infty,\ ErrorforK \leftarrow \infty$
3:   $OutputWeights \leftarrow 1$ // All d dimensions have same weight initially
4:     **for** each $k=$ low to high **do**
5:       $ErrorforK \leftarrow 0$
6:       $ClosestNeighbors \leftarrow GetNeighbors(data, k, T)$
7:       $DimensionalRegressor(T)$ // Algorithm 2
8:       **if** $MinimumError > Errorfork$ **then**
9:         $MinimumError \leftarrow Errorfork$
10:         $OutputWeight \leftarrow W_T$
11:       **end if**
12:     **end for**
13: **return** $OutputWeight$
14: **end procedure**

Algorithm 2

1: **procedure** DIMENSIONAL REGRESSION
2:    **for** each i = 1 to d **do** // d is the number of dimensions
3:        $Line_i \leftarrow ConstructLine(ClosestNeighbors, i)$ // As discussed in Sect. 4
4:        $PredictedTestVal_i \leftarrow T_i * a + b$
5:        $E_i \leftarrow (PredictedTestVal_i - ActualTestVal_i)^2$
6:    **end for**
7:    **if** $\forall\ i\ E_i$ is equal **then**
8:        $W_T \leftarrow 1$
9:    **else**
10:        **for** each i = 1 to d **do**
11:            $weight_i \leftarrow max(E_i)/E_i$
12:            $W_T \leftarrow weight_i$
13:        **end for**
14:    **end if**
15: $Errorfork \leftarrow \sum_{j=1}^{d} E_j$
16: **return** W_T
17: **end procedure**

It would be appropriate to state that a lower error value in predicting the line indicates that the constructed regression line is optimal in nature and fits the neighborhood of test data. Hence we conclude that the value of dependent variable predicted via the line of best fit is approximately correct and thus a higher weight should be assigned for a more optimal line or we can say a line with lower squared error. This intuition is captured by assigning weights in inverse proportion to the error in prediction for this dimension, hence a feature with high error value is assigned lower weight and the feature with lower error value is assigned higher weight. The squared error in prediction of neighbors (line 15 of Algorithm 2) is computed and stored in $Errorfork$. A lower value of the squared error indicates that the weight values chosen using the nearest neighbors are appropriate. We then select the value of the parameter k for which the calculated error is minimum and hence assigns the corresponding weight vector W_T (line 8–10 of Algorithm 1). On this weighted feature vector we evaluate the anomaly score via a scaled Manhattan distance metric as discussed in the section below. The approach demonstrated in Algorithms 1 and 2 is a completely novel idea for dimension wise assigning weights in inverse proportion to error.

5.3 Neural Network for Anomaly Detection

After the weights have been assigned to the feature vector via our proposed algorithm, we calculate the anomaly scores as described by [12] for evaluating our model. For calculating anomaly score we are using a simple feed-forward neural network with a input layer with the size of our feature vectors and one hidden layer of 200 dimensions. After experimenting with different number of neurons in the hidden layer, we found out that results are best reported at 200 neurons. All the layers are fully connected. The higher size of hidden layer introduces sparsity in our network and helps in capturing the inter-feature relations which might be present. Following subsections explain other building blocks of neural network. We later discuss ablation studies for each in Table 2. We define the loss by the negative log-likelihood function which maximized the probability that sample gets classified as user or impostor. Learning is done through back-propagation of the losses through our network [10] (Fig. 2).

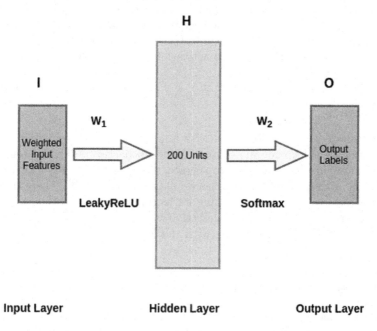

Fig. 2. Illustration of neural network architecture. Here **W1** and **W2** represents the weight matrices which are our parameters to be learnt. **H** is the hidden layers of size 200. **I** and **O** are input and output layers respectively.

5.4 Dropout

We use dropout [20] after our hidden layer which act as a regularizer and restricts over-fitting. During our training stage we randomly delete the nodes of each hidden layer with a certain probability p for each input sample. These neurons do not participate in the back-propagation learning. In testing time, the weights are correspondingly divided by $1-p$. Using dropout, forces the rest of the neurons in the hidden layers to learn more robust features and depend lesser on other specific neurons. In [20] more details are provided which show that using dropout can be an economic alternative to ensembling various network architectures.

5.5 Batch Normalization

After every fully-connected layer, we use batch normalization [11] before the respective activation functions. Using batch normalization we monitored the gap between training and testing loss over epochs narrowed down. This led to better generalization.

5.6   Leaky ReLU

Non-linear function Rectified linear unit (ReLU) is preferred to sigmoid or hyperbolic-tan because it simplifies backpropagation, makes learning faster while also avoiding saturation. However for large gradients, ReLU [16] can cause particular neurons to die and not participate in learning at-all. LeakyReLU's have a small positive gradient $f(z) = max(0.01x, x)$ which prevent this dying of a neuron. We applied Leaky ReLU as our activation function after the fully connected layers.

5.7   Adam

In recent times, several algorithms (with implemented software tools) are available for training a deep neural network. While stochastic gradient descent (SGD) for quite some time have been the top choice, there has been study which indicate some of the obvious flaws [14] in the vanilla implementation. There have been some attempts to automatically tune its learning rate thus resulting in much faster convergence. For anomaly detection we use Adam [13] instead of SGD which required a lot of fine-tuning with the learning rate and over 500 epochs to converge.

5.8   Inputs to Neural Network

The dataset consists of keystroke timing information of 51 users, where each user is made to type **.tie5Roanl** as password. All the 51 users enrolled for this data collection task typed the same password in 8 different sessions with 50 repetitions per session thus making each user to type 400 times in total (Table 1).

Table 1. Hyperparameters used in our neural network

Name	Specification
Dropout	0.3
LeakyReLU	0.01
Adam	0.001
Epochs	200

6   Experimental Setup

In this section we discuss the experimental setup, evaluation criteria used and the performance of our proposed model. We evaluated our approach on the CMU keystroke dynamics benchmark dataset [12] where *51* users were designated for this task. We demonstrate the effectiveness of our model with the average equal error rate (EER). We compare the results with various proposed anomaly detectors/classifiers which have been used in literature. We used the python library Keras [4] for building our neural network architecture. All our experiments were carried out on a Pentium 4th generation machine with 4 GB memory. For experiments, we took the same 200 initial timing feature vector per-user as before. 10% of training data was kept aside as validation data for hyperparameter tuning.

6.1   Training

We frame keystroke dynamics based authentication as a one-class classification problem. For authentication, neural network learns one model per user, rejects anomalies to the learned model as impostors, and accepts inliers as the genuine user. Consider a scenario in which a users long-time password has been compromised by an impostor. We assume the user to be practiced in typing their own password, while the impostor is unfamiliar with it (e.g., typing it for the first time). We measure how well each of our detectors is able to discriminate between the impostors typing and the genuine users typing in this scenario. We start by designating one of our 51 subjects as the genuine user, and the rest as impostors. We train an anomaly detector by extracting 200 initial timing feature vectors for a genuine user from the dataset. We repeat this process, designating each of the other subjects as the genuine user in turn thus creating models equal to number of distinct subjects or users. Unlike most existing approaches, which only use actual users data at training time, our model leverage data from background users to enhance the models discriminative capability thus improving the prediction performance. We randomly took 5 samples from each background users as negative samples. Note that these 5 random samples were carefully chosen such that no impostor samples that were used in testing were shown during the time of training. For this problem setting, we use the evaluation criteria as mentioned in [12] in order to have comparison on same grounds.

6.2   Testing

We take last 200 passwords typed by the genuine user from the dataset. These 200 timing feature vectors acts as test data. Scores generated in this step acts as the user scores. Next, we take initial 5 passwords typed by each of the 50 impostors (i.e., all subjects other than the genuine user) from the dataset which acts as the impostor scores. Thus we form a test dataset of 200 positive samples and 250 negative samples per user which we provide to our neural network and record the output predictions. If s denotes the predictions, the corresponding anomaly score was calculated as $1 - s$ [12]. Intuitively, if s is close to 1.0, the test vector is similar to the training vectors, and with s close to 0.0, it is dissimilar (Fig. 3).

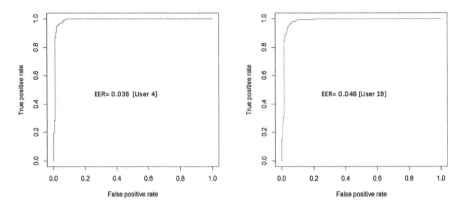

Fig. 3. Shows ROC curve for different users with their equal error rate (EER) value where user number corresponds to the user as labeled in CMU dataset.

6.3   Empirical Evaluation for User Authentication

Based on the genuine user scores and impostor scores generated in the steps above, we generate the ROC curve for the actual (genuine) user. Then we calculate the equal error rate from the ROC curve where the equal error rate corresponds to that point on the curve where the false positive rate (false-alarm) and false negative rate (miss) are equal. We repeat the above four steps, in total of 51 times where every time each of the subsequent user is taken as the genuine user from the 51 distinct users in turn, and calculate the equal-error rate for each of the genuine users. Finally we compute the mean of all 51 equal-error rates which gives us the performance value for all users, and the standard deviation which will give us the measure of its variance across subjects. In order to ensure comparison on same grounds we have used exactly the same evaluation criteria as stated by [12] on our proposed approach. The train-test data split was also kept the same.

Table 2. Ablation study of our neural network model showcasing the increase of performance for each additional component we use. Note that for this test we used tan-h activation when we opt not to use LeakyReLU. All components are added independently on the base model. Accuracy is based on the multiclass user recognition problem.

Model architecture	EER	Time taken (min)
Base model	0.0701	1.54
With Dropout	0.0674	3.18
With LReLU	0.0623	4.53
With Batch-Norm	0.0602	5.12
Our 3 layer model	**0.051**	5.81

7   Results

Table 3 shows the comparison of 16 other proposed keystroke timing algorithms with our proposed approach. Comparison is shown with 16 other algorithms which used the same dataset and the same evaluation criteria thus assuring an objective comparison. Our model is able to achieve an average equal error rate (EER) of **0.051** and with a standard deviation (stddev) of 0.042 across 51 subjects. The average equal error rate (EER) shown in the Table 3 are the fractional rates between 0.0 and 1.0, not the percentages. Clearly from Table 3, our model performs superior than other proposed techniques in comparison.

Table 3. Comparison of 16 different keystroke timing pattern algorithms that uses the same CMU keystroke timing dataset and evaluation criteria in terms of average equal error rate (EER) (with standard deviation shown in brackets).

Model/algorithm	Average EER (stddev)	Source
Our proposed algorithm (with 2 new engineered features)	**0.051(0.040)**	
Our proposed algorithm (without 2 new engineered features)	**0.054(0.042)**	
Median vector proximity	0.080(0.055)	[2]
Manhattan-Mahalanobis (no outlier)	0.084(0.056)	[23]
Manhattan-Mahalanobis (outlier)	0.087(0.060)	[23]
Manhattan (scaled)	0.0962(0.0694)	[12]
Nearest neighbor (Mahalanobis)	0.0996(0.0642)	[12]
Outlier count (z-score)	0.1022(0.0767)	[12]
SVM (one-class)	0.1025(0.0650)	[12]
Mahalanobis	0.1101(0.0645)	[12]
Manhattan (filter)	0.1360(0.0828)	[12]
Neural network (auto-assoc)	0.1614(0.0797)	[12]
Euclidean	0.1706(0.0952)	[12]
Fuzzy logic	0.2213(0.1051)	[12]
k Means	0.3722(0.1391)	[12]
Neural network (Standard)	0.8283(0.1483)	[12]

8   Conclusion and Future Work

In this paper we investigate the problem of authenticating users based on keystroke timing pattern. We engineered new features namely bigram time and inversion ratio time apart from the features already given in the CMU keystroke timing dataset. Besides engineering new features we also proposed a simple and robust nearest neighbor based regression algorithm. We evaluated our results and compared it against 14 other algorithms that used the same dataset and evaluation criteria thus providing performance comparison on equal grounds. Although simple, it proved to be effective as it outperformed competing algorithms as shown in Table 3.

Future work involves extending our work for soft keys or touch pad keys and in addition to timing pattern features we can use users pressure patterns as well in order to authenticate users. We are planning to experiment with different curve fitting techniques as well. We plan on extending our models to other available datasets on this domain. We would also like to investigate if transfer learning can help with user authentication and identification for large pool of users when trained from a limited dataset.

Acknowledgement. This work was supported by http://metabolomics.iiit.ac.in/ and we would like to thank them for their support.

References

1. Dvorak, A., Merrick, N., Dealey, W., Ford, G.: Typewriting behavior (1936)
2. Al-Jarrah, M.M.: An anomaly detector for keystroke dynamics based on medians vector proximity. J. Emerg. Trends Comput. Inf. Sci. **3**(6), 988–993 (2012)
3. Bergadano, F., Gunetti, D., Picardi, C.: User authentication through keystroke dynamics. ACM Trans. Inf. Syst. Secur. (TISSEC) **5**(4), 367–397 (2002)
4. Chollet, F.: Keras (2015). https://github.com/fchollet/keras
5. Cover, T., Hart, P.: Nearest neighbor pattern classification. IEEE Trans. Inf. Theory **13**(1), 21–27 (1967)
6. Forsen, G.E., Nelson, M.R., Staron Jr., R.J.: Personal attributes authentication techniques. Technical report, DTIC Document (1977)
7. Gaines, R.S., Lisowski, W., Press, S.J., Shapiro, N.: Authentication by keystroke timing: some preliminary results. Technical report, DTIC Document (1980)
8. Giot, R., Hemery, B., Rosenberger, C.: Low cost and usable multimodal biometric system based on keystroke dynamics and 2D face recognition. In: ICPR (2010)
9. Gunetti, D., Picardi, C.: Keystroke analysis of free text. ACM Trans. Inf. Syst. Secur. (TISSEC) **8**(3), 312–347 (2005)
10. Hecht-Nielsen, R.: Theory of the backpropagation neural network. In: International Joint Conference on Neural Networks, IJCNN, vol. 1, pp. 593–605 (1989)
11. Ioffe, S., Szegedy, C.: Batch normalization: accelerating deep network training by reducing internal covariate shift. In: ICML (2015)
12. Killourhy, K.S., Maxion, R.A.: Comparing anomaly-detection algorithms for keystroke dynamics. In: DSN (2009)
13. Kingma, D.P., Ba, J.: Adam: a method for stochastic optimization (2014). CoRR abs/1412.6980
14. Le, Q.V., Ngiam, J., Coates, A., Lahiri, A., Prochnow, B., Ng, A.Y.: On optimization methods for deep learning. In: ICML (2011)
15. Leggett, J., Williams, G.: Verifying identity via keystroke characterstics. Int. J. Man Mach. Stud. **28**(1), 67–76 (1988)
16. Maas, A.L., Hannun, A.Y., Ng, A.Y.: Rectifier nonlinearities improve neural network acoustic models (2013)

17. Monrose, F., Rubin, A.D.: Keystroke dynamics as a biometric for authentication. Future Gener. Comput. Syst. **16**(4), 351–359 (2000)
18. Peacock, A., Ke, X., Wilkerson, M.: Typing patterns: a key to user identification. IEEE Secur. Priv. **2**(5), 40–47 (2004)
19. Schapire, R.E.: A brief introduction to boosting. In: IJCAI (1999)
20. Srivastava, N., Hinton, G.E., Krizhevsky, A., Sutskever, I., Salakhutdinov, R.: Dropout: a simple way to prevent neural networks from overfitting. J. Mach. Learn. Res. **15**, 1929–1958 (2014)
21. Syed, Z., Banerjee, S., Cheng, Q., Cukic, B.: Effects of user habituation in keystroke dynamics on password security policy. In: HASE (2011)
22. Syed, Z., Banerjee, S., Cukic, B.: Leveraging variations in event sequences in keystroke-dynamics authentication systems. In: HASE (2014)
23. Zhong, Y., Deng, Y., Jain, A.K.: Keystroke dynamics for user authentication. In: CVPR (2012)

Applications

HypGraphs: An Approach for Analysis and Assessment of Graph-Based and Sequential Hypotheses

Martin Atzmueller[1]([⊠]), Andreas Schmidt[1], Benjamin Kloepper[2],
and David Arnu[3]

[1] Research Center for Information System Design, University of Kassel,
Kassel, Germany
{atzmueller,schmidt}@cs.uni-kassel.de
[2] ABB Corporate Research Center Germany, Ladenburg, Germany
benjamin.kloepper@de.abb.com
[3] RapidMiner GmbH, Dortmund, Germany
darnu@rapidminer.com

Abstract. The analysis of sequential patterns is a prominent research topic. In this paper, we provide a formalization of a graph-based approach, such that a directed weighted graph/network can be extended using a sequential state transformation function, that "interprets" the network in order to model state transition matrices. We exemplify the approach for deriving such interpretations, in order to assess these and according hypotheses in an industrial application context. Specifically, we present and discuss results of applying the proposed approach for topology and anomaly analytics in a large-scale real-world sensor-network.

1 Introduction

The analysis of sequential patterns, e.g., as a sequence of states, is a prominent research topic with broad applicability, ranging from exploring mobility patterns [7] to technical applications [9]. The DASHTrails approach [7] provides a comprehensive modeling approach for comparing hypotheses with such sequences (trails), in order to identify those hypotheses that show the largest evidence concerning the observed data.

This paper presents the HypGraphs analysis approach (extending DASH-Trails) for analyzing and assessing sequential hypotheses in the form of transition matrices given a directed weighted network. The application context is given by (abstracted) alarm sequences in industrial production plants in an Industry 4.0 context. Specifically, we consider the analysis of the plant topology and anomaly detection in alarm logs. The assessment of the static structure can help in identifying problems in the setup of the production plant, while dynamic relations can be applied for the analysis of unexpected (critical) situations. Our contribution is summarized as follows:

© Springer International Publishing AG 2017
A. Appice et al. (Eds.): NFMCP 2016, LNAI 10312, pp. 231–247, 2017.
DOI: 10.1007/978-3-319-61461-8_15

1. We outline a flexible analytics approach for assessing graph-based and sequential hypotheses in a *graph interpretation* of weight-attributed directed networks.
2. For that, we show how to embed the recent DASHTrails [7] approach for distribution-adapted *modeling and analysis* of sequential hypotheses and trails. Furthermore, we motivate and show the advantages of this state-of-the-art Bayesian approach compared to a typically applied frequentist approach for testing network associations.
3. Furthermore, we outline the application of the proposed approach in an industrial context, for the analysis of plant *structures* in industrial production contexts, as well as for detecting *anomaly indicators* concerning disrupting dynamic processes.

The remainder of the paper is structured as follows: Sect. 2 discusses related work. After that, Sect. 3 outlines the proposed method. Next, Sect. 4 presents results of a case study of HYPGRAPHS in the industrial context. Finally, Sect. 5 concludes with a discussion and outlines interesting directions for future work.

2   Related Work

The investigation of sequential patterns and sequential trails are interesting and challenging tasks in data mining and network science, in particular in graph mining and social network analysis. A general view on modeling and mining of ubiquitous and social multi-relational data is given in [5] focusing on social interaction networks. Orman et al. [18] defines a sequence-based representation of networks. Then the sequential patterns are used to characterize communities. For comparing hypotheses and sequential trails, the HypTrails [20] algorithm has been applied to sequential (human) navigational trails derived from web data. In [7] we have presented the DASHTrails approach that incorporates probability distributions for deriving transitions. Extending the latter, the proposed HYPGRAPHS framework provides a more general modeling approach. Using general weight-attributed network representations, we can infer transition matrices as *graph interpretations*, while HYPGRAPHS consequently also relies on Markov chain modeling [15,21] and Bayesian inference [21,22].

Sequential pattern analysis has also been performed in the context of alarm management systems, where sequences are represented by the order of alarm notifications. Folmer et al. [11] proposed an algorithm for discovering temporal alarm dependencies based on conditional probabilities in an adjustable time window. To reduce the number of alarms in alarm floods, Abele et al. [3] performed root cause analysis with a Bayesian network approach and compared different methods for learning the network probabilities. Vogel-Heuser et al. [23] proposed a pattern-based algorithm for identifying causal dependencies in the alarm logs, which can be used to aggregate alarm information and therefore reduce the load of information for the operator. In contrast to those approaches, the proposed approach is not only about detecting sequential patterns. We provide a systematic approach for the analysis of (derived) sequential transition matrices and their comparison relative to a set of hypotheses. Thus, similar to evidence networks in the context of social networks, e.g., [17], we model transitions assuming a certain interpretation of the data towards a sequential representation.

The detection and analysis of anomalies and outliers [12] in network-structured data is a novel research area, e.g., for identifying new and/or emerging behavior, or for identifying detrimental or malicious activities. The former can be used for deriving new information and knowledge from the data, for identifying events in time or space, or for identifying interesting, important or exceptional groups [4, 19]. In contrast to approaches for anomaly detection that only provide a classification of anomalous and normal events, we can assess different anomaly hypotheses: applying the proposed approach, we can then generate an anomaly indicator – as a potential kind of second opinion method, e.g., for assessing the state of a production plant that can help for indicating explanations and traces of unusual alarm sequences. Then, using the network representation, we can analyze anomalous episodes relative to structural (plant topology) as well as dynamic (alarm sequence) episodes.

3   Method

In the following section, we first provide an overview on the proposed approach. After that, we discuss the modeling and analysis steps in detail.

3.1   Overview

We start with a set of directed weighted networks. Then, we interpret these weights for constructing transitions between states (denoted by the nodes of the network) and compare this *data* to *hypotheses* that can also be constructed using the network-based formalizations. Adapting the modeling principles of the DASHTrails approach that we have presented in [7] to our network formalism, we model transition matrices given a probability distribution of certain states. We assume a discrete set of such states Ω corresponding to the nodes of the network (without loss of generality $\Omega = \{1, \ldots, n\}$, $n \in \mathbb{N}, |\Omega| = n$). Then, assuming a certain *network interpretation* of the weights of the edges, we construct transitions between states. As shown in Fig. 1, we perform the three following steps, that we discuss below in more detail:

1. Modeling: Determine a transition model given the respective weighted network using a *transition modeling function* $\tau : \Omega \times \Omega \to \mathbb{R}$. Transitions between sequential states $i, j \in \Omega$ are captured by the elements m_{ij} of the transition matrix M, i.e., $m_{ij} = \tau(i, j)$. Then, we collect sequential transition matrices for the given network (data) and hypotheses.
2. Estimation: Apply HypTrails, cf. [20] on the given data transition matrix and the respective hypotheses, and return the resulting evidence.
3. Analysis: Present the results for semi-automatic introspection and analysis, e.g., by visualizing the network as a heatmap or characteristic sequence of nodes.

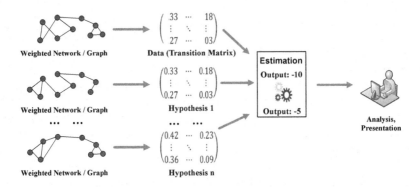

Fig. 1. Overview on the HYPGRAPHS modeling and analysis process.

3.2   Modeling and Comparing Graph-Based Network Interpretations

As outlined above, we derive the transition matrices (modeling transitions between states) using a certain *transition modeling function* $\tau : \Omega \times \Omega \rightarrow \mathbb{R}$, as described below. The transition modeling function τ captures a certain interpretation of these weights. In the case of hypotheses, these correspond to link traversal probabilities from one state to another state, represented by the respective individual nodes. Equivalently, we can represent the obtained directed and weighted graph in the form of an adjacency matrix, where the individual values of an entry (i, j) correspond to the weight of the link between nodes i and j; as an hypothesis this can be interpreted as a transition probability between two states i and j.

Modeling. For modeling, we consider a sequential interpretation (according to the first order Markov property) of the original data with respect to the obtained transition probabilities (Markov chain). Thus, using τ, we can model (derived) transition matrices corresponding to the *observed data*, e.g., given frequencies of alarms on measurement points, as well as hypotheses on sequences of alarms. For data transition matrices, we need to map the transitions into derived counts in relation to the data; for hypotheses we provide the (normalized) transition probabilities. That is, for hypothesis testing, we can directly convert the weighted network using the defined transition modeling function (i.e., we convert the obtained values to probabilities by row-normalization).

For observed sequences, we can simply construct transition matrices counting the transitions between the individual states, e.g., corresponding to the set of alarms. Then, $\tau(i, j) = |suc(i, j)|$, where $suc(i, j)$ denotes the successive sequences from state i to state j contained in the sequence. For deriving transition matrices from a probability distribution over certain events, for example, we need to apply a more complex modeling approach. We refer to [7, 20] for more details on modeling and inference, respectively.

Assessment. For assessing a set of hypotheses that consider different transition probabilities between the respective states, we apply the core Bayesian estimation step of HypTrails [20] for comparing a set of hypotheses representing beliefs about transitions between states. In summary, we utilize Bayesian inference on a first-order Markov chain model. As an input, we provide a (data) matrix, containing the transitional information (frequencies) of transition between the respective states, according to the (observed) data. In addition, we utilize a set of hypotheses given by (row-normalized) stochastic matrices, modeling the given hypotheses. The estimation method outputs a set of evidence values, for the set of hypotheses, that can be used for ranking these. Also, using the evidence values, we can compare the hypotheses in terms of their significance.

Specifically, hypotheses are expressed in terms of belief in Markov transitions, such that we distinguish between common and uncommon transitions between the respective states. Then, for each hypothesis, we construct the belief matrix for subsequent inference. Given the data (matrix), we elicit a conjugate Dirichlet prior and finally obtain the evidence using marginal likelihood estimation. Here, the evidence denotes the probability of the data given a specific hypothesis. Thus, this can also be interpreted as the relative plausibility of a hypothesis. Then, the hypotheses can be ranked in terms of their evidence.

Furthermore, a central aspect of the method is an additional parameter (k) indicating the *belief* in a given hypothesis: the higher the value of is k the higher is the belief in the respective hypothesis matrix, i.e., its parameter configuration. Given a lower value of k, the hypothesis is assigned more tolerance, such that other (but similar) parameter configurations become more probable. Then, for assessing a hypothesis, we monitor its performance with increasing k, typically relative to the data itself (as a kind of upper bound), the uniform hypothesis (as a random baseline) and competing hypotheses.

In contrast, the quadratic assignment procedure [14] (QAP) is a frequentist approach for comparing network structures. For comparing two graphs G_1 and G_2, it estimates the correlation of the respective adjacency matrices [14] and tests a given graph level statistic, e.g., the graph covariance, against a QAP null hypothesis. QAP compares the observed graph correlation of (G_1, G_2) to the distribution of the respective resulting correlation scores obtained on repeated random row and column permutations of the adjacency matrix of G_2. As a result, we obtain a correlation value and a statistical significance level according to the randomized distribution scores.

As we will show in our experiments below, the applied Bayesian inference technique has significant advantages compared to the typically applied frequentist approach for comparing networks based on graph correlation using the QAP test [14]: we not only know whether a hypothesis is significantly correlated with the data, but we can also compare hypotheses (and their significance) relative to each other (given Bayes factor analysis, cf. [13]). In particular, this also holds for those hypotheses that are not correlated with the data, obtaining a total ranking for likely and unlikely hypotheses. Furthermore, we can express our *belief* in the hypothesis relative to the data, and analyze the impact of that on the evidence concerning the likelihood estimate.

4   Case Study

Below, we first outline our application context and discuss the instantiation of the proposed approach. After that, we discuss the collected datasets before we describe results of a case study of HYPGRAPHS in the industrial context in detail.

4.1   Application Context

In many industrial areas, production facilities have reached a high level of automation nowadays. Here, knowledge about the production process is crucial, targeting both static relations like the topological structure of a plant and the modeling of operator notifications (alarms), and dynamic relations like unexpected (critical) situations. Assessment of the static structure can help in identifying problems in the setup of the production plant. The dynamic relations involve analytics for supporting the operators, e.g., for diagnosis of a certain problem. The objective of the BMBF funded research project "Early detection and decision support for critical situations in production environments"[1] (short FEE) is to detect critical situations in production environments as early as possible and to support the facility operator with a warning or even a recommendation on how to handle this particular situation. The consortium of the FEE project consists of several partners including also application partners from the chemical industry. These partners provide use cases for the project and background knowledge about the production process, which is important for designing analytical methods. In this paper, we utilize HYPGRAPHS in this application context, both for static (topology) and dynamic (alarm log) analysis.

4.2   HYPGRAPHS Instantiation

In an industrial production plant, alarms for certain measurement points occur if the value of the measurement is not within a specified value range. Therefore, by intuition, an alarm sequence (for a given point in time, or interval) represents an abstracted state of the production plant. Then, we can utilize the "normal" long running state of the plant as the "normal behavior", excluding known anomalous episodes.

We perform two kinds of analyses. First, we compare the normal behavior to the overall topology of the plant, i.e., corresponding to transitions between different functional units of the plant. Second, we compare the normal behavior to our anomaly hypotheses, which are defined by the captured anomalies. Doing that, we assume that the sequence of alarms indicates a certain normal or abnormal (process) behavior. We can then compare the (historic) long running state of the plant to the current state for obtaining indicators about possible normal or abnormal situations.

---

[1] http://www.fee-projekt.de.

4.3   Dataset

In our experiments, we used a dataset from the FEE project that was collected in a petrochemical plant; it includes a variety of data from different sources such as sensor data, alarm logs, engineering- and asset data, data from the process information management system as well as unstructured data extracted from operation journals and instructions.

We used alarm logs for a period of two months as well as Piping and Instrumentation Diagrams (P&IDs) [10] which represent the topological structure of the facility, i.e., capturing the piping of the considered petro-chemical process along with installed equipment (pumps, valves, heat-exchangers, etc.) and instrumentation used to control the process. P&IDs are usually composed of several sub-diagrams with disjoint system elements. Connections between elements on different P&IDs are captured in textual form at the corresponding pipe or other connecting elements. Commonly, the structuring of P&IDs follows in some way the structure of the captured process and plant capturing different areas. In our dataset, the titles of P&IDs suggest such a structuring of the P&IDs around major equipment like tanks, reactors, processing columns, etc. (e.g. 'Input vessel - desorption plant', 'Preheater - desorption plant', 'Desorber - desorption plant', 'Steam/condensate - auxiliary materials'). We also used text data from the operation journals to verify anomalous events. The characteristics of the applied real-world dataset are shown in Tables 1 and 2. According to standards [1,2] P&IDs are used to identify the measurements (temperatures, flows, level, pressures, etc.) in the process, using identifiers of the respective measurement points with up to 5 letters. The alarms in the alarms logs are defined based on measurements captured in the P&ID diagrams, usually as a threshold value on the corresponding measurements; the entries in the alarm log reference the measurements in the P&IDs by a matching identifier.

Table 1. Characteristics of the real-world dataset (petrochemical plant) for a period of two months

	Count
Anomalies	4
P&IDs	63
P&IDs referenced in alarm log	55
Alarms referencing measurement points in P&IDs	59.623
Distinct alarms referencing P&IDs	327
P&ID transitions (between distinct P&IDs)	384
Topological connections (between distinct P&IDs)	299

4.4   Matrix Construction

Before constructing the transition matrices, we first identified anomalous events by looking at the operation journals. We used this background knowledge to divide the dataset into nine disjoint time slots with five normal and four abnormal episodes. For abnormal episodes, we empirically determined a time window of one hour spanning the anomalous event starting half an hour before the event and ending half an hour after the event. In practice, the length of this time window is a parameter that needs to be determined according to application requirements. All nine time slots together covered the whole time (two months). Note that we only used the alarms that could be mapped to a P&ID. The distribution of alarms and P&IDs for the different time slots is shown in Table 2.

Table 2. Overview on normal/abnormal episodes for the real-world dataset (petrochemical plant)

#	Episode	#Alarms	#Distinct alarms	#Distinct P&IDs
1	Normal1	10503	66	34
2	Abnormal1	86	12	9
3	Normal2	8382	91	31
4	Abnormal2	212	14	5
5	Normal3	6130	74	31
6	Abnormal3	220	17	7
7	Normal4	6318	89	29
8	Abnormal4	1516	127	30
9	Normal5	26256	278	44

For each time slot, we constructed a transition matrix M by counting the consecutive transitions in the sequence of the alarm log. Formally, let $A = <a_1, a_2, ..., a_n>$ be a sequence of alarms which represents the alarm log. We created a function, which maps alarms to P&IDs $\text{map}(a_t)$ and retrieved the P&IDs contained in the alarm log $P = \{\text{map}(a_t)|a_t \in A\}$. Then, the weights m_{ij} for the $|P| \times |P|$ transition matrix M are given by the number of transitions from p_i to p_j with $(p_i, p_j) \in P \times P$:

$$m_{ij} = |\{(a_t, a_{t+1}), a_t, a_{t+1} \in A, \text{map}(a_t) = p_i, \text{map}(a_{t+1}) = p_j\}|$$

For the data matrix corresponding to the alarm data, we can then just utilize the obtained count data, denoting the number of transitions between the states. For creating hypotheses, we normalize the data by row in order to obtain a stochastic matrix.

We also extracted data from the P&IDs corresponding to the plant organization in terms of functional units. As described above each P&ID corresponds to such a functional unit, containing several sensors that can then trigger respective alarms if the corresponding measurements are not within a specified value range. A P&ID shows the process and instrumentation structure and also links to other P&IDs with respect to certain flows (material, energy, information) that connects the process structure. Given the P&IDs in PDF format, we converted the data to XML and extracted the necessary information for modeling all possible (directed) links between the individual P&IDs in a network-based representation of the overall plant modeling.

4.5   Results and Discussion

According to our hypotheses, we expect that the functional units of the plant also model functional dependencies as observed by alarm sequences. Furthermore, we expect, that normal episodes (sequences) should be "close" to the normal (long running) behavior. Accordingly, abnormal sequences should be "away" from the normal (reference) behavior – in terms of evidence. As we will see below, we can confirm these hypotheses using Bayes factor analysis [13]. As a baseline, we furthermore apply the presented QAP method. Since a (data) transition matrix should be explained best by its corresponding hypothesis, we constructed a respective row-normalized data transition matrix. In addition, we constructed a uniform hypotheses (square matrix, all entries being 1) as a random baseline. A good hypothesis explaining the normal behavior should be between both, however, relatively close to the data.

Topological Analysis. As previously discussed, the document structure of P&IDs capture to a certain extent the structure of the process plant they describe. Simply put, the designer of the P&IDs decided to put elements on the same diagram because they are closely related (although, sometimes graph layout consideration might override this rule of thumb). Consequently, the measurements captured on a P&ID are more closely related to measurements across different P&IDs. Since measurements are used to define alarm messages, it seems a valid assumption that consequently alarms in the alarm logs should reference measurements on the same P&ID with a higher probability than measurements from different P&IDs. Based on this assumption, we formulated our first hypothesis to test HypGraphs on the industrial dataset: For topological analysis, we utilized the given P&ID graph containing directed links between the P&IDs. We checked whether the alarm sequences (normal behavior) can be explained by a uniform topology model, where we assume that transitions between all linked P&IDs are equally likely. The results are shown in Fig. 2. We observe that the uniform topology hypothesis does not explain the data well since it is significantly away (larger k) compared to the data and close to the random baseline. In contrast, an "encapsulated topology" hypothesis fits the data relatively well, assuming that transitions in alarm sequences mainly occur local to the specific P&IDs. This confirms our expectations and indicates a good performance of plant and alarm management in general.

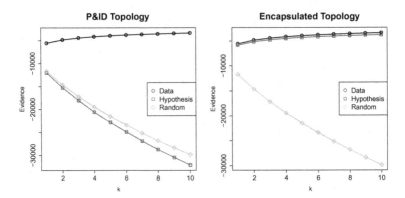

Fig. 2. Topological analysis: uniform topology hypothesis and local topology hypothesis.

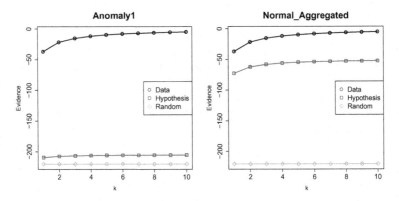

Fig. 3. Artificial local topology baseline: example of an anomalous and a normal hypothesis.

Furthermore, we double-checked the data against an artificial baseline, assuming only transitions local to P&IDs (in that case, the transition matrix becomes a diagonal matrix). Results are shown in Fig. 3. We observed strict differences between normal and abnormal episodes, two examples are shown in the figure: While abnormal situations are far away from the local hypothesis, normal situations are significantly closer, but these, cannot "explain" only local transitions, indicating that most transitions but not all conform to this artificial situation. We also checked the rankings of the normal and abnormal episodes comparing the respective hypotheses to the real data (normal behavior) and the artificial local topology baseline. Using Kendall's-Tau as a correlation measure (0.61), the ranking was not very consistent, indicating that the local topology assumption alone is too simple in order to be explainable by the observed data.

Overall, we observe that we can verify structural modeling assumptions using HYPGRAPHS (given in the P&ID structure) using the collected data from the alarm logs. We already observe distinct differences between abnormal and normal episodes.

Anomaly Analytics. In the start phase of the FEE project, a series of workshops and interviews were executed for identifying potential Big Data and analytics applications. One of the identified analytics tasks was anomaly detection. The idea behind that application scenario is that retrospective analysis of disrupting events often uncovers that a situation could have been handled better, if the operators or process experts had been involved earlier and would have been pointed to the relevant data. Thus, we developed a description of the current and desired situation to identify the right analytics questions:

- **Current Situation:**
  - *Who*: Operator in the operating room, shift leader (in the operating room), process engineer, process manager (in the office).
  - *What*: Anomalies (e.g., uncommon oscillations) in a plant need to be recognized as early as possible. If such cases are not recognized by the operator, serious problems can occur (product is not usable, unplanned plant shutdown, etc.) and staff with higher expertise need to be informed.
  - Challenge: Anomalies are not easy to detect manually. New technologies like advanced controllers make anomalies even more difficult to detect. Furthermore, operators usually inform an expert when a problem has occurred and they are not able to handle it. In addition, diagnostics of an anomaly by process engineers and managers is usually time-consuming.
- **Desired Situation:**
  - System: informs the operator about a possible anomaly. The operator performs an analysis and diagnosis of the situation and informs the expert.
  - Expert: automated updates about possible anomalies; can track long term trends.
  - Users: pointed to relevant measurements for supporting diagnostic activities.

In the context of anomaly analytics, our results indicate the significance of the proposed HypGraphs approach for specifically supporting analysis and diagnosis tasks.

In particular, for anomaly analysis of the alarm data, we used the partitioning of the dataset into normal and abnormal episodes. Then, we checked both abnormal and normal situations against the assumed "normal behavior" of the plant that is observed for the long running continuous process. In the analysis, we applied a typical estimation procedure using separate training and tests sets, such that the data and the tested hypotheses do not overlap in time. However, since we have only had data covering a two months period available we also tested the hypotheses against the aggregated normal behavior covering all normal episodes. It turned out, that the findings reported below are also consistent across these different evaluation periods; we observe the same (significant) trends, confirming the individual results even on larger scale.

Figure 4 shows the different anomaly hypotheses corresponding to the different anomaly episodes (cf. Table 2). We observe that the anomalies are well distinguishable (using Bayes factor analysis [13]). The anomalies are "well away" from data (more than factor 3 for higher k), indicating a significant deviation

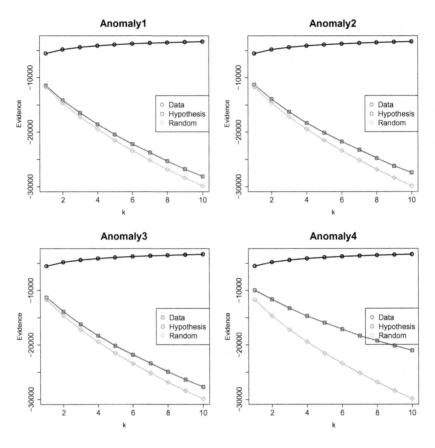

Fig. 4. Normal behavior (data) compared to different anomaly episodes (anomaly 1–4) and a random baseline (uniform hypothesis).

from the data. Furthermore, we observe distinct characteristics of the anomalies, observing the trends with increasing k. Anomalies 1–3 are of the same class and show similar characteristics, while Anomaly 4 conforms to another real-world class of a disrupting event, also showing different characteristics in terms of evidence. We also performed an analysis using the QAP procedure for the anomaly data, correlating the transition matrices corresponding to the normal behavior and the abnormal episodes. These results support the findings of the Bayesian approach, showing a correlation close to zero that was not significant. However, while confirming the deviation, QAP does not allow to derive a (significance-based) ranking of the different hypotheses here, in contrast to our proposed approach.

Figure 5 shows results of comparing exemplary normal episodes (as hypotheses) with the normal behavior (data) – the results for the rest of the normal episodes show equivalent trends. We observe significant differences compared to the anomaly hypotheses. Using the Bayes factors technique, we also observe that

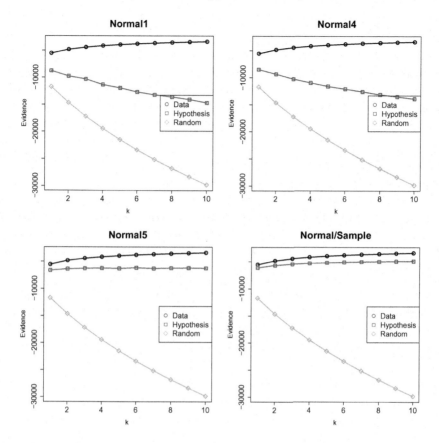

Fig. 5. Normal behavior (data) compared to different normal episodes and a random baseline (uniform hypothesis).

the normal behavior is well detectable, the hypotheses are sufficiently "close" to the data hypotheses. In addition, we also compared shorter normal periods (using random samples of the normal behavior) in order to exclude control for the different sizes of the alarm distributions. The bottom right chart of Fig. 5 shows an example - the findings confirm our results for the other episodes well.

For the normal episodes, we also applied QAP analysis, using the graph correlation measure on transition matrices corresponding to the normal behavior and the respective normal episodes described above. Here, we observed significant ($p = 0.01$) correlation values between 0.42 and 0.72, with a ranking of the normal hypotheses that is consistent with the Bayesian approach. In essence, this suggests that our findings are rather robust against the selected statistical measure.

Retrospective as well as realtime analysis can be supported, for example, by according visualization approaches summarizing anomalous episodes in the form of heatmaps, or by directly tracing anomalous sequences on a detailed level of analysis. Figure 6 demonstrates an example of a heatmap visualization, showing data for the (aggregated) anomalies 1–3 compared to the long term behavior of the data: Rows/columns of the matrix refer to the individual P&IDs. The aggregations refer to different types of real-world anomalies and we can observe distinct "fingerprints" of the transitional episodes. Then, by inspecting the different cells (corresponding to transitions of alarms between a pair of P&IDs), the respective data points (sequences of alarms) can be assessed in detail, e.g., showing the corresponding alarm messages or sensor reading. Please note, that this visualization can be applied for static data, i.e., for retrospective analysis, as well as for dynamic analysis, e.g., utilizing a suitable time window for data aggregation on the current (alarm) log data stream.

In summary, these analysis results indicate the significance of the HypGraphs approach for anomaly analytics, concerning detection, analysis and diagnosis tasks. Applying HypGraphs we can compare different hypotheses to the "normal behavior" and identify normal and abnormal episodes in a data-driven way. In contrast to typical frequentist approaches like QAP, we can obtain a ranking of both the normal and abnormal episodes, enabling a comprehensive view on the data for anomaly analytics, complemented by suitable visualizations. Furthermore, there are several visualization options to be used for dashboards,

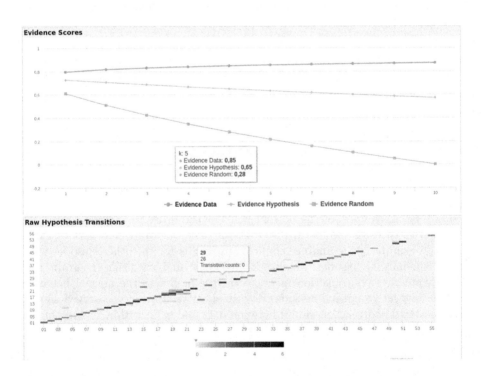

Fig. 6. Example of a dashboard with a heatmap visualization, showing "fingerprints" for the long term behavior and for an anomalous sequence. Rows/columns refer to the individual P&IDs.

e.g., the obtained evidence plots, using extended heatmaps, or a detailed view on sequences of nodes corresponding to individual alarms.

4.6 Big Data Aspects

With time periods longer than two months or with very detailed sensor readings, the amount of data can quickly get overwhelming for normal computation systems. In this case, a distributed storage and computation system can handle the requirements of evaluating several years of production data. The RapidMiner [16] platform, for example, can be integrated with Hadoop systems such that preprocessing and analytical processes built on a local machine can be transferred to the big data environment. In the context of the FEE project, we target a two layered architecture where long running and computationally expensive processes run in the Hadoop infrastructure and either the prepared data or the final models, in this case the transition matrix M, can be applied on a local machine. The computation can be executed, e.g., in a Spark/MapReduce [8] process and the orchestration and deployment can be handled with RapidMiner, for which the HypGraphs approach is already implemented as an independent extension.[2]

5 Conclusions

This paper outlined the HypGraphs approach for modeling and comparing graph-based and sequential hypotheses using first-order Markov chain models. Our application context is given by structural and anomaly analytics in Industry 4.0 contexts, i.e., of (abstracted) alarm sequences in industrial production plants. We applied a real-world dataset in an Industry 4.0 context, specifically in the scope of the FEE project.

In summary, we considered the analysis of the plant topology and anomaly analytics in alarm logs, which was identified as one major application in the project. Our results indicate that the proposed HypGraphs approach is well suited for analyzing and assessing the transition networks, respectively the corresponding alarm sequences. We could identify distinct differences between abnormal and normal episodes, e.g., in order to derive an anomaly indicator. We also verified the modeling of plant topology and alarm setup. The results can help for analysis and inspection of the corresponding alarm sequences, e.g., for detailed analysis and diagnosis of anomalies. This enabled directly the inspection, for example, of a deviating sequence through a drill-down into the data. Furthermore, results can be transparently visualized, e.g., in the form of heatmaps, and embedded into Big Data dashboards.

For future work, we aim to extend the analysis using high diversity data, i.e., with longer time periods, different event and anomaly settings. We are also

---

[2] https://github.com/rapidminer/rapidminer-extension-hypgraphs.

investigating options for detecting descriptive anomaly patterns [6]. Furthermore, including more background knowledge on known relations on plant configuration and the extension to an unsupervised approach for anomaly detection is another interesting direction.

Acknowledgements. This work was funded by the BMBF project FEE under grant number 01IS14006. We wish to thank Leon Urbas (TU Dresden) and Florian Lemmerich (GESIS, Cologne) for helpful discussions, also concerning Florian's implementation of HypTrails (https://bitbucket.org/florian_lemmerich/hyptrails4j) [20].

References

1. ANSI/ISA-S51.1-1979 (R1993): Process instrumentation terminology
2. ISO 14617-6:2002 Graphical Symbols for Diagrams - Part 6: Measurement and Control Functions
3. Abele, L., Anic, M., Gutmann, T., Folmer, J., Kleinsteuber, M., Vogel-Heuser, B.: Combining knowledge modeling and machine learning for alarm root cause analysis. In: MIM, pp. 1843–1848. International Federation of Automatic Control (2013)
4. Akoglu, L., Tong, H., Koutra, D.: Graph based anomaly detection and description. Data Min. Knowl. Disc. **29**(3), 626–688 (2015)
5. Atzmueller, M.: Data mining on social interaction networks. JDMDH **1** (2014)
6. Atzmueller, M.: Detecting community patterns capturing exceptional link trails. In: Proceedings of IEEE/ACM ASONAM. IEEE Press, Boston (2016)
7. Atzmueller, M., Schmidt, A., Kibanov, M.: DASHTrails: an approach for modeling and analysis of distribution-adapted sequential hypotheses and trails. In: Proceedings of WWW 2016 (Companion). IW3C2/ACM (2016)
8. Becker, M., Mewes, H., Hotho, A., Dimitrov, D., Lemmerich, F., Strohmaier, M.: SparkTrails: a MapReduce implementation of HypTrails for comparing hypotheses about human trails. In: Proceedings of WWW (Companion). ACM, New York (2016)
9. Buddhakulsomsiri, J., Zakarian, A.: Sequential pattern mining algorithm for automotive warranty data. Comput. Ind. Eng. **57**(1), 137–147 (2009)
10. Cook, R.: Interpreting piping and instrumentation diagrams. Blog-Entry, September 2010. http://www.aiche.org/chenected/2010/09/interpreting-piping-and-instrumentation-diagrams
11. Folmer, J., Schuricht, F., Vogel-Heuser, B.: Detection of temporal dependencies in alarm time series of industrial plants. In: Proceedings of 19th IFAC World Congress, pp. 24–29 (2014)
12. Hawkins, D.: Identification of Outliers. Chapman and Hall, London (1980)
13. Kass, R.E., Raftery, A.E.: Bayes factors. J. Am. Stat. Assoc. **90**(430), 773–795 (1995)
14. Krackhardt, D.: QAP partialling as a test of spuriousness. Soc. Netw. **9**, 171–186 (1987)
15. Lempel, R., Moran, S.: The stochastic approach for link-structure analysis (SALSA) and the TKC effect. Comput. Netw. **33**(1), 387–401 (2000)
16. Mierswa, I., Wurst, M., Klinkenberg, R., Scholz, M., Euler, T.: YALE: rapid prototyping for complex data mining tasks. In: Proceedings of KDD, pp. 935–940. ACM, New York (2006)

17. Mitzlaff, F., Atzmueller, M., Benz, D., Hotho, A., Stumme, G.: Community assessment using evidence networks. In: Atzmueller, M., Hotho, A., Strohmaier, M., Chin, A. (eds.) MSM/MUSE -2010. LNCS, vol. 6904, pp. 79–98. Springer, Heidelberg (2011). doi:10.1007/978-3-642-23599-3_5

18. Orman, G.K., Labatut, V., Plantevit, M., Boulicaut, J.F.: A method for characterizing communities in dynamic attributed complex networks. In: Proceedings of IEEE/ACM International Conference on Advances in Social Networks Analysis and Mining, pp. 481–484 (2014)

19. Ranshous, S., Shen, S., Koutra, D., Harenberg, S., Faloutsos, C., Samatova, N.F.: Anomaly detection in dynamic networks: a survey. WIREs: Comput. Stat. **7**(3), 223–247 (2015)

20. Singer, P., Helic, D., Hotho, A., Strohmaier, M.: HypTrails: a Bayesian approach for comparing hypotheses about human trails. In: Proceedings of WWW. ACM, New York (2015)

21. Singer, P., Helic, D., Taraghi, B., Strohmaier, M.: Memory and structure in human navigation patterns. PLoS ONE **9**(7) (2014)

22. Strelioff, C.C., Crutchfield, J.P., Hübler, A.W.: Inferring Markov chains: Bayesian estimation, model comparison, entropy rate, and out-of-class modeling. Phys. Rev. E **76**(1), 011106 (2007)

23. Vogel-Heuser, B., Schütz, D., Folmer, J.: Criteria-based alarm flood pattern recognition using historical data from automated production systems (aPS). Mechatronics **31** (2015)

Mining Chess Playing as a Complex Process

Stefano Ferilli[1,2(✉)] and Sergio Angelastro[1]

[1] Dipartimento di Informatica, Università di Bari, Bari, Italy
{stefano.ferilli,sergio.angelastro}@uniba.it
[2] Centro Interdipartimentale di Logica e Applicazioni, Università di Bari, Bari, Italy

Abstract. The main objective of this paper is checking whether, and to what extent, advanced process mining techniques can support efficient and effective knowledge discovery in complex domains. This is done on chess playing, cast as a process. A secondary objective is checking whether the discovered information can provide interesting insight in the game rules and strategies, and/or may support effective game playing in future matches. Experimental results provide a positive answer to the former question, and encouraging clues on the latter.

1 Introduction

The increasing complexity of the processes that underlie most human activities nowadays motivated the research on automatic process mining [14] and managing. The WoMan framework for process management was shown in [5,6] to be able to handle efficiently and effectively very complex processes where other state-of-the-art systems fail, especially due to short loops, duplicate activities and large concurrency (4–5 activities). Motivated by these results, the main objective of this paper is to stress WoMan and check whether it is able to handle even much more complex processes, this way confirming its power. We pursue our objective considering the domain of chess playing, that we cast as a (complex) process. Since WoMan is a general framework, it does not need any specific tailoring for handling this new domain. More specifically, in our vision, playing a chess match corresponds to enacting a process. We consider a task as the occupation of a specific square of the chessboard by a specific kind of piece (e.g., "black rook in a8"), and the involved resources as the two players: 'white' and 'black'. Matches are initialized by starting 32 activities corresponding to the initial positions of all pieces on the chessboard. Each move terminates some activities (i.e., removes pieces from the squares they occupy) and starts new activities (i.e., occupies some squares by pieces). This setting represents a tough testbed to evaluate the process mining state-of-the-art due to the following features:

- very high concurrency (the number of pieces on the chessboard during most of the match is in the order of dozens, which is beyond the reach of many current process mining systems [6]);
- huge number of tasks (in principle, 752 possible combinations piece-square);

© Springer International Publishing AG 2017
A. Appice et al. (Eds.): NFMCP 2016, LNAI 10312, pp. 248–262, 2017.
DOI: 10.1007/978-3-319-61461-8_16

- very huge number of transitions (moves are in the order of 10^{123}, and we may have different transitions for the same move, depending on whether it is a take or not); and
- extremely huge number of cases (estimated to be $10^{10^{50}}$).

It also features short and nested loops (a piece going back on a square after a number of moves), optional activities (a given move may or may not take an opponent's piece) and duplicate tasks (a piece may occupy the same square at different stages of the match, and thus in different contexts).

WoMan adopts the declarative setting [11], working in First-Order Logic (FOL). This is a further advantage in this domain, since various kinds of relationships involved in chess playing are fundamental to correctly grasp high-level information about it. Based on these premises, a secondary objective emerged: that is, to check whether the learned information provides useful hints to humans about chess playing, using both qualitative and quantitative evaluations. However, it is important to point out that we are not interested in assessing the superiority of declarative approaches to process mining versus more classical ones in general, nor in directly comparing WoMan to other systems.

This paper is organized as follows. After summarizing basics and related work on Process Mining, we briefly recall the architecture and formalism of WoMan, and we present some considerations about the issues and opportunities raised by the formalism, which is more powerful than those usually adopted in the Process Mining literature. Section 5 deals with experimental settings and results, showing that WoMan is able to learn this complex kind of process and to discover interesting information in that domain. Finally, Sect. 6 draws some conclusions and outlines future work issues.

2   Basics and Related Work

A *process* consists of a suitable combination of actions, performed by any kind of agents [1,2]. It may involve sequential, concurrent, conditional, or iterative execution [12]. A *workflow* is a formal specification of a process [12]. A *case* is a particular execution of actions compliant to a given workflow. Cases can be described by *traces*, consisting of lists of events associated to *steps* (time points). Several traces may be collected in *logs* [13]. A *task* is a generic piece of work, defined to be executed for many cases of the same type. An *activity* is the actual execution of a task by a *resource* (an agent that can carry it out). A workflow can be modeled as a directed graph, where Nodes are associated to states or activities, while Edges represent the potential flow of control among activities; they can be labeled with probabilities and/or boolean conditions on the state of the process, which determine whether they will be traversed or not [1]. In literature have been proposed, for representing processes, Finite State Machines (FSMs) [3] and Hidden Markov Models (HMMs); both are unsuitable to model concurrency. Subsequent works have mainly focused on Petri nets, or on their restriction WorkFlow nets (WF-nets) [12], purposely developed to express the control flow in a process. E.g., [12,14] learn models in the form of WF-nets. The α-algorithm family [4,13,15] mines processes in the class of sound Structured WF-nets, which can handle parallelism only between pairs of tasks. WoMan framework describes process models with a FOL-based approach. This setting is called Declarative Process Mining and is recognized as being very important when dealing with particularly complex models and domains [11]. Control flow is modeled by imposing only a set of constraints that must be satisfied when executing the process activities. The chess domain is far from the purposes of classical process mining techniques, indeed nothing is done in the literature. Others approaches dealt with this domain and recent work on Machine Learning applied to chess playing investigated the use of Deep Learning [8] and Neural Networks [10]. However, the proposed approaches are sub-symbolic (i.e., models are not human readable) and require huge amounts of examples. Thanks to its declarative approach, WoMan overcomes both these shortcomings.

3 Advanced Process Mining with WoMan

The WoMan framework [5,6] introduced some important novelties and peculiarities in the process mining and management landscape. A fundamental one is the pervasive use of FOL as a representation formalism, that allows to describe contextual information using relationships. In particular, it works in the Logic Programming fragment of FOL [9]. Due to space limitations, the interested reader is referred to [5,6] for a detailed presentation of WoMan's architecture, features and representation formalism. Here, we will briefly and intuitively recall the notions that will be useful to understand the proposed application.

3.1 Architecture

Several modules are included in WoMan to perform the different tasks involved in learning and managing process models. **WIND** (Workflow INDucer) allows one to learn or refine a process model according to a case, after the case events are completely acquired. The refinement may affect the structure and/or the probabilities. Differently from all previous approaches in the literature, WoMan's learning module is *fully incremental*: not only can it refine an existing model according to new cases whenever they become available, it can even start learning from an empty model and a single case, while others need a (large) number of cases to draw significant statistics before learning starts. This is a significant advance with respect to the state-of-the-art, because continuous adaptation of the learned model to the actual practice can be carried out efficiently, effectively and transparently to the users [5].

PAT (Process Analysis Tool) allows one to perform several kinds of analysis on the learned model, such as identifying the most frequently used model components, comparing the frequent components among different models, comparing the structure of different models, etc.

The supervision module allows one to check whether new cases are compliant with a given model. **WEST** (Workflow Enactment Supervisor and Trainer) takes the case events as long as they are available, and returns information about their compliance with the currently available model for the process they refer to. The output for each event can be 'ok', 'error' (e.g., when closing activities that had never begun, or ending the execution while activities are still running), or a set of warnings denoting different kinds of deviations from the model (e.g., unexpected task or transition, preconditions not fulfilled, unexpected resource running a given activity, etc.).

While in supervision mode, the prediction modules allow one to foresee which activities the user is likely to perform next, or to understand which process is being carried out among a given set of candidates. Specifically, **SNAP** (Suggester of Next Action in Process) hypothesizes which are the possible/appropriate next activities that can be expected given the current intermediate status of a process execution, ranked by confidence. Confidence here is not to be interpreted in the mathematical sense; it is determined based on a heuristic combination of several parameters associated with the history of the current partial process execution. Finally, given a case of an unknown workflow, **WoGue** (Workflow Guesser) returns a ranking (by confidence) of a set of candidate process models.

3.2 Representation

Following the foundational literature [1, 7], WoMan takes as input trace elements consisting of 7-tuples, represented as logic atoms of the form:

$$\texttt{entry}(T, E, W, P, A, O, R).$$

where T is the event timestamp, E is the type of the event (one of 'begin_process', 'end_process', 'begin_activity', 'end_activity'), W is the name of the workflow the process refers to, P is a unique identifier for each case, A is the name of the activity, O is the progressive number of occurrence of that activity in that case, and R is an optional field can be used to specify the resource that is carrying out the activity (it was not present in [1, 7]).

A model describes the structure of a workflow using the following elements:

tasks: the kinds of activities that are allowed in the process;
transitions: the allowed connections between activities.

The core of the model is the set of transitions, since they carry all the information about the flow of activities during process execution. A transition can be formalized as

$$t : I \Rightarrow O$$

where t is a unique identifier and I, O are multisets of tasks. t is enabled if all input tasks in I are concurrently active. It occurs when, after stopping (in any order) the concurrent execution of all tasks in I, the concurrent execution of all output tasks in O is started (again, in any order). Given this behavior, activities that terminate play the same role as *tokens* in Petri Nets. Carrying on this analogy, the current set of terminated activities not yet used to fire any transition is the *marking*, and firing a transition *consumes* all the tokens reported in the corresponding multiset I.

Each task or transition t is associated to the multiset C_t of training cases in which it occurred (a multiset because a task or transition may occur several times in the same case). It can be exploited both for guiding the conformance check of new cases and for computing statistics on the use of tasks and transitions. In particular, it allows us to compute the probability of occurrence of a task/transition t in a model learned from n training cases as the relative frequency $|C_t|/n$. Tasks and transitions can also specify the involved resources (i.e., for transitions, the agents that must carry out the activities in I and O).

As shown in [5,6], this representation formalism is more powerful than Petri or Workflow Nets [14], that are the current standard in Process Mining. It can smoothly express complex task combinations and models involving invisible or duplicate tasks, which are problematic for those formalisms.

4   Supervision and Prediction Issues

The increased power of WoMan's representation formalism for workflow models raises some issues that must be tackled. In Petri Nets, since a single graph is used to represent the allowed flow(s) of activities in a process, at any moment in time during a process enactment, the supervisor knows exactly which tokens are available in which places, and thus which tasks are enabled. So, the prediction of the next activities that may be carried out is quite obvious, and checking the compliance of a new activity with the model means just checking that the associated task is in the set of enabled ones. Conversely, in WoMan the activity flow model is split into several 'transitions', and different transitions may share some input and output activities, which allows them to be composed in different ways with each other. As a consequence, many transitions may be eligible for application at any moment, and when a new activity takes place there may be some ambiguity about which one is actually being fired. Such an ambiguity can be resolved only at a later time. Let us see this through an example.

Example 1. Suppose that the given model includes the following transitions:

$$t_1: \{x\} \Rightarrow \{a,b\} \quad t_2: \{x,y\} \Rightarrow \{a\} \quad t_3: \{w\} \Rightarrow \{d,a\}$$

and that the current marking (i.e., the set of the latest activities that were terminated in the current process enactment, but not yet used to fire any transition) is $\{x,y,z\}$. Now, suppose that activity a is started. It might indicate that either transition t_1 or transition t_2 have been fired. Also, if an activity d is currently being executed due to transition t_3, the current execution of a might correspond to the other output activity of transition t_3, which we are still waiting to happen to complete that transition. Each of these options would change in a different way the process evolution, as follows:

t_1: firing this transition would consume x, yielding the new marking $\{y, z\}$ and causing the system to wait for a later activation of b;

t_2: firing this transition would consume x and y, yielding the new marking $\{z\}$ and causing the completion of transition t_2;

t_3: firing this transition would not consume any element in the marking, but would cause the completion of transition t_3.

We call each of these alternatives a *status*. So, given an intermediate status of the process enactment and a new activity that is started, there may be different combinations of transitions that are compliant with the new activity, and one may not know which is the correct one until a later time. Thus, all the corresponding alternate evolutions of the status must be carried on by the system. When the next event is considered, each of these evolutions is a possible status of the process enactment. On the one hand, it poses again the same ambiguity issues; on the other hand, it may point out that some current alternate statuses were wrong. So, as long as the process enactment proceeds, the set of alternate statuses that are compliant with the activities carried out so far can be both expanded with new branches, and pruned of all alternatives that become incompatible with the activities carried out so far.

Note also that each alternative may be compliant with a different set of training cases, and may rise different warnings (in the previous example, one option might be fully compliant with the model, another might rise a warning for task preconditions not fulfilled, and the other might rise a warning for an unexpected agent running that activity). WEST takes note of the warnings for each alternative and carries them on, because they might reflect secondary deviations from the model that one is willing to accept. Wrong alternatives will be removed when they will be found out to be inconsistent with later events in the process enactment. So, the question arises about how to represent each alternative status. As suggested by the previous example, we may see each status as a 5-tuple

$$\langle M, R, C, T, W \rangle$$

recording the following information:

M   the marking, i.e., the set of terminated activities that have not been used yet to fire a transition, each associated with the agent that carried it out and to the transition in which it was involved as an output activity;

R   the set of activities that are 'ready' to start, i.e., the output activities of transitions that have been fired in the status, and that the system is waiting for in order to complete those transitions;

C   the set of training cases that are compliant with that status;

T   the set of (hypothesized) transitions that have been fired to reach that status;

W   the set of warnings raised by the various events that led to that status.

The system also needs to remember, at any moment in time, the set *Running* of currently running activities and the list *Transitions* of transitions actually carried out so far in the case. The set of statuses is maintained by WEST, as long as the events in a case are processed.

As regards predictions, we recall that, due to the discussed set of alternate statuses that are compliant with the activities carried out at any moment, differently from Petri Nets it is not obvious to determine which are the next activities that will be carried out. Indeed, any status might be associated to different expected evolutions. The good news is that, having several alternate statuses, we can compute statistics on the expected activities in the different statuses, and use these statistics to determine a ranking of those that most likely will be carried out next.

Confidence of activity predictions carried out by SNAP is determined based on a heuristic combination of several parameters associated with the possible alternate process statuses that are compliant with the current partial process execution. Specifically, the activities that can be carried out next in a given status are those included in the *Ready* component of that status, or those belonging to the output set of transitions that are enabled by the *Marking* component of that status. The status parameters used for the predictions are the following:

1. frequency of activities across the various statuses (activities that appear in more statuses are more likely to be carried out next);
2. number of cases with which each status is compliant (activities expected in the statuses supported by more training cases are more likely to be carried out next);
3. number of warnings raised by the status (activities expected in statuses that raised less warnings are more likely to be carried out next);
4. confidence of the tasks and transitions as computed by the multiset of cases supporting them in the model (activities supported by more confidence, or belonging to transitions that are associated to more confidence, are more likely to be carried out next).

Also process prediction carried out by WoGue is based on the possible alternate statuses identified by WEST when applying the events of the current process enactment to the candidate models. In this case, the candidate models are ranked by decreasing performance of their 'best' status, i.e. the status reporting best performance in (one or a combination of) the above parameters.

5   Evaluation: Chess Playing as Process Execution

In the following, we evaluate the performance of the WoMan's activity prediction approach in the chess playing domain. This domain is appropriate because it is characterized by much more variability and subjectivity in the users' behavior, and it does not involve a 'correct' underlying model, just some kind of 'typicality' can be expected. The chess domain also allows us to evaluate the performance of the process prediction approach.

For our experiments, we downloaded from the Website of the Italian Chess Federation (http://scacchi.qnet.it) 400 reports of actual top-level matches played by Anatolij Karpov and Garry Kasparov (200 matches each), and translated them from PGN (Portable Game Notation), an open source format representing the moves in algebraic notation, to the input format of WoMan.

As already pointed out, playing a chess match corresponds to enacting a process. A task is the occupation of a specific square of the chessboard by a certain piece and the corresponding activities are characterized by the time at which the piece starts occupying a square and the time at which it leaves that square. Each task/activity is encoded as a 4-characters string denoting, respectively: chessboard file, chessboard rank, piece (in the following we used the Italian initials: p = pawn, t = rook, a = bishop, c = knight, d = queen, r = king), and player (b = white, n = black). Transitions correspond to moves: indeed, each move of a player (resource) terminates some activities (since it moves pieces away from the squares they currently occupy) and starts new activities (that is, the occupation by pieces of their destination squares). Three kinds of transitions are allowed: normal moves, takes and castlings. Note that a transition just reports the initial and final piece positions, without any specification of the game constraints that allow it to be a valid move (e.g., no pieces on the pathway, no own piece in destination square, etc.).

Example 2. The following are valid examples of chess-based tasks/activities:

$b4pb$ means that a white pawn is in square $b4$;
$c8rn$ means that the black king is in square $c8$.

On the other hand, the following are valid examples of chess-based transitions on some activities:

t_j: $\{c6cn\} \Rightarrow \{e5cn\}$ a black knight moves from square $c6$ to square $e5$ without taking any piece;
t_k: $\{b4pb, c3an\} \Rightarrow \{b4an\}$ a black bishop moves from square $c3$ to square $b4$ and takes a white pawn (indeed, activity $c3an$ is terminated, indicating that the pawn is removed from square $b4$);
t_l: $\{e1rb, a1tb\} \Rightarrow \{c1rb, d1tb\}$ castling: the white king moves two squares left from its initial square $e1$, and a white rook goes from its initial square $a1$ to the square immediately right of the white king.

Each log entry bears the following information:

T a progressive number indicating the event timestamp;
E one of the allowed event types:
    begin_of_process   the start of a match;
    begin_of_activity   a certain piece is placed in a certain square;
    end_of_activity   a certain piece is removed from a certain square;
    end_of_process   the end of a match.
W the name of the process model the entry is referred to;
P a unique match identifier, obtained by concatenation of the following data: name of white player, name of black player, place and date of the match;

Table 1. Dataset statistics

Player	# Matches				Events		Tasks		Runtime (sec)	
	Total	White	Black	Draw	Total	Avg	Total	Avg	Total	Avg
Karpov	200	79	48	73	45088	225.44	22344	111.72	43.192	0.215
Kasparov	200	79	39	82	45244	226.22	22422	112.11	41.192	0.205
Total	400	158	87	155	90332	225.83	44366	110.915	94.4	0.236

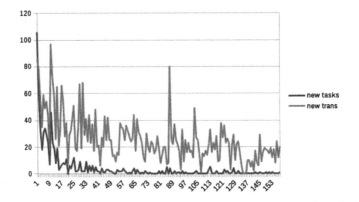

Fig. 1. Incremental learning behavior for white (Color figure online)

A the name of the activity;
O the progressive number of occurrence of A in P
R the player (*white* or *black*) responsible for the beginning or end of activity.

As regards the processes of interest, we considered three processes, corresponding to the possible match outcomes: *white* wins, *black* wins, or *draw*. Then we used the following WoMan functionality. First we used WEST, to check compliance of each training case with the current models, in combination with WIND, to learn and refine the models after the compliance check of each case. On the final learned models, we used PAT, to see whether interesting information was discovered and 'compiled' in the models. Finally, we used SNAP and WEST to check whether correct predictions could be obtained using the learned models, in this way indirectly evaluating the quality of the discovered information.

Table 1 reports statistics on the dataset and on the learning runtime of WoMan. The number of matches for the three processes is similar for the two players. Note that the average (*avg*) number of events and tasks are nearly identical, suggesting that it does not depend on the specific players. Also the average runtime needed to learn the models is almost identical for the two subsets, suggesting it only depends on the number of cases. For the whole dataset, including twice as much cases as the single subsets, average runtime is still comparable to those of the subsets, which confirms that WoMan has an acceptable time complexity (linear or loglinear) in the number of processed cases. Less than a quarter of a second is needed on average to incorporate a new match in the

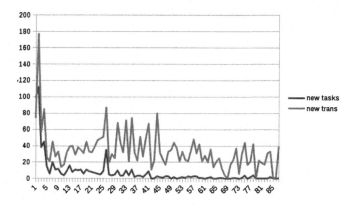

Fig. 2. Incremental learning behavior for black (Color figure online)

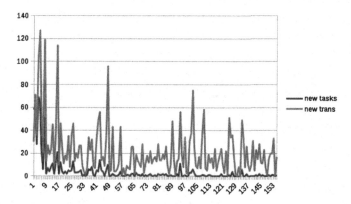

Fig. 3. Incremental learning behavior for draw (Color figure online)

available model, which means, considering that each match involves nearly 226 events on average, that each event is processed in about 1 ms. We may conclude that WoMan has proven its ability to learn such complex kinds of workflows. Figures 1, 2 and 3 show the dynamic learning behavior. The graphics report the number of new tasks (blue line) and transitions (red line) that were found in each new training case, compared to those already learned from all previous cases. The peaks, especially for tasks, become progressively lower and sparser, which confirms that the system is actually converging on a subset of very relevant tasks and transitions.

Table 2 shows that the number of tasks and transitions is comparable in the models learned from the two training subsets. The number of tasks is almost the same also in the overall dataset, which was expected because the number of all possible piece positions is small (752) compared to the number of tasks in the training sets. Nevertheless, the models actually use only up to 90% of all possible tasks, suggesting that some piece positions are undesirable. Conversely, the number of different transitions from the subsets to the overall dataset grows,

Table 2. Model statistics

	Karpov			Kasparov			Karpov + Kasparov		
	White	Black	Draw	White	Black	Draw	White	Black	Draw
# Matches	79	48	73	79	39	82	158	87	155
# Tasks (avg)	629	624	588	635	567	615	681	663	658
	3.14	3.12	2.94	3.17	2.83	3.07	1.7	1.65	1.39
# Transitions (avg)	2665	2055	1949	2445	1575	2318	4083	3006	3434
	13.32	10.27	9.74	12.22	7.87	11.59	10.20	7.51	8.58

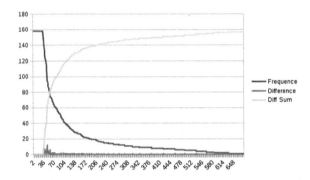

Fig. 4. Statistics on tasks frequency for the *white* process (Color figure online)

as expected, but the growth is less than linear, even if they are much less than all possible ones. This may indicate that only a small portion of moves is really relevant to players. We also note that the number of different moves is neatly larger when the white wins than when the black wins.

Let us now focus on the frequency of tasks and transitions. Figures 4 and 5 refer to the *white* process, but it is representative of *black* and *draw*, as well. So, we will not report the graphics for these processes. The blue line plots the number of cases in which the different tasks (a) and transitions (b) occur in the model, by decreasing frequency. The same information is shown also as difference between adjacent values (red line), and as cumulative difference in occurrences (yellow line). The initial plateau in (a) is given by the initial position of pieces on the chessboard, that of course occurs in all training cases. The shape shows a clear drop in frequency very early (i.e., toward the left) along the x-axis, that is evident also in the series of close peaks in the red line. It suggests which are the most frequent items on which further analysis can be carried out. Table 3 shows the 20 most frequent (non-initial) tasks for each process (in decreasing order from left to right). The frequency of task/transition t refers to the probability of occurrence of t ($|C_t|/n$) in its model. Very interestingly, we find in the top positions the tasks that correspond to well-known openings widely acknowledged in the chess literature. E.g., in all processes *d4pb* (white pawn in d4) is in the top

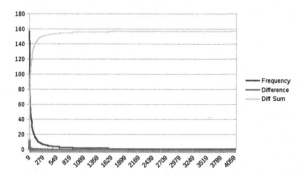

Fig. 5. Statistics on transitions frequency for the *white* process (Color figure online)

Table 3. Most frequent tasks for the different processes

	1	2	3	4	5	6	7	8	9	10
w	d4pb	f6cn	f3cb	f1tb	e4pb	g1rb	f8tn	g8rn	c3cb	c5pn
b	f6cn	f3cb	d4pb	g8rn	f8tn	c3cb	g1rb	f1tb	e4pb	c5pn
d	f6cn	d4pb	f8tn	f3cb	g8rn	f1tb	g1rb	c3cb	c5pn	c4pb
	11	12	13	14	15	16	17	18	19	20
w	c4pb	e6pn	d5pn	c6cn	d6pn	d1tb	d7cn	f4pb	g6pn	e5pn
b	c4pb	d5pn	c6cn	e6pn	e5pn	d1tb	d6pn	d4pn	d5pb	b5pn
d	e6pn	e7an	e4pb	d5pn	c6cn	d5pb	d6pn	e5pn	g3pb	d4pn

3 positions, *f3cb* (white knight in f3) is in the top 4 positions, and *e4pb* (white pawn in e4) is in the top 13 positions.

Since many tasks and transitions appear in all processes, the question arises about which ones, if any, are more characteristic or discriminant for the final outcomes of a match (i.e., for each process). We took the set of k most frequent tasks for each process, for different values of k, and computed the differences among such sets. The lower k, the more characteristic the items in such a difference are expected to be for a given process. E.g., considering the top 10 tasks in Table 3, *e4pb* (white pawn in e4) is present in *white* and *black*, but not in *draw*, suggesting that it may be discriminant for the latter. Conversely, *c4pb* (white pawn in c4) somehow characterizes *draw*, because it is in its 10 most frequent tasks, but not in the top 10 of the others. Clearly, there is an obvious skew of frequencies in favor of initial moves, where the possibilities are more limited, and thus we expect to find initial moves of the pieces for low values of k, while for higher values of k we expect that all kinds of processes have exploited all possible initial moves, and thus the characterizing or discriminant items are more related to the middle of the match. Table 4 shows, for each process, all the tasks that *never* occurred (but that occurred in the other processes), and all those that *always* occurred (but never occurred in the other processes).

Table 4. Characteristic and discriminant tasks for the different processes

White	Never	a4cn, a5an, b2pn, b2rb, b4cb, b7cb, c1cn, c8ab, d2pn, e2cn, e8cb, f1cn, f1dn, f2cn, f2dn, f3tn, f4an, g1dn, g1tn, g2dn, g2tb, h1dn, h2an, h3db, h7ab
	Always	a1cn, a7cn, a7pb, a7rn, a8cn, a8rn, b2rn, b7pb, b7rb, b8ab, c2pn, c2rn, c5rb, d7rb, d8cb, e7pb, e7rb, f7ab, f7rb, g1ab, g1rn, g2pn, g2rn, g3rn, g8an, h1cb, h2db, h2pn, h2tb, h3rn, h5rb, h7cb, h8cn, h8db, h8dn
Black	Never	a1db, a2cb, a2cn, a4rn, a5rb, a5rn, a6ab, a6db, a6pb, a6rb, a7ab, a7tn, b1db, b3tb, b4tn, b5rb, b6rb, b7cn, b8an, b8rn, c4rb, c6rb, c8cn, c8db, d1an, d2an, d3rn, d3tn, d6rb, d7ab, d8rn, d8tb, e1dn, e2an, e4rb, e5rb, e6db, e8ab, e8dn, f6cn, f6db, f6rb, f6tb, f7cn, f7dn, f7pb, f7tn, f8cb, f8cn, g2cb, g3an, g4cb, g4tb, g5an, g5cn, g5db, g6cb, g6db, g7dn, g7pb, h2ab, h2dn, h3pn, h4cn, h4rn, h4tb, h5db, h5rn, h6cb, h6db, h6dn, h6tn, h7db, h7dn, h7tn
	Always	a1ab, a1cb, a1rb, a1tb, a2rb, a3cn, a3rb, c1rn, c3rn, d2rn, e1cn, f3rn, f4rn, g1an, g5rb, g8ab, h1db
Draw	Never	a2db, a4rb, a5ab, a6cb, b1an, b3dn, b4rb, b6db, b6pb, b8cb, c7pb, c7rn, d1cn, d4rn, e2pn, e3pn, e6ab, f2pn, f3cn, f4tn, f5rb, f5rn, f5tb, f8ab, g2tn, h1tn, h2tn, h3tn, h4tn, h5ab, h8an, h8tb
	Always	a1an, a2an, a6rn, a8ab, b3an, b4rn, b5cn, b5rn, c1an, c1cb, c4rn, c5rn, d8cn, e1ab, e1an, e2rn, e3rn, f1an, g1db, g8db, h4an, h7pb

These considerations are confirmed for transitions. For $k = 20$, characterizing moves for white-winning matches are *[f2pb]-[f4pb]* (white pawn moving two squares from its initial position f2) and *[b8cn]-[d7cn]* (black knight moving from b8 to d7). The latter move may be interesting, because it cannot be done before the pawn initially placed in d7 moves away, and thus there are less chances that it happens frequently in the initial moves.

As regards the prediction of next activities, we report in Table 5 some statistics obtained from an 80%–20% training-test set random split procedure for the different processes. Column 'predictions' reports in how many cases SNAP returned a prediction (when tasks or transitions not present in the model are executed in the current enactment, WoMan assumes a new kind of process is enacted, and avoids making predictions). Among the cases in which WoMan makes a prediction, column 'correct' reports in how many of those predictions the correct activity (i.e., the activity that is actually carried out next) is present, and column 'Ranking' reports how close it is to the first element of the ranking (1.0 meaning it is the first in the ranking, possibly with other activities, and 0.0 meaning it is the last in the ranking). The 'Quality' index is the product of these values:

$$Quality = Predictions \cdot Correct \cdot Ranking$$

It ranges from 0.0, meaning that predictions are completely unreliable, to 1.0, meaning that WoMan always makes a correct prediction. Interestingly, when a prediction is made it is almost certainly correct, and contains the next actual activity at the top of the ranking. The number of predictions made is also interesting, given that in more than half of the match WoMan is able to suggest which will be the next action in the game. This is noteworthy, also compared to the state-of-the-art [8,10], and indicates that significant information about chess was discovered in the models. The worse performance is on 'black', which is the one with less training cases.

Table 5. Prediction statistics

	Activity				Process				
	Predictions	Correct	Ranking	Quality	Acc	Avg	F	A	L
Black	0.42	0.98	1.0	0.41	0.50	2.0	0.25	14.62	77.12
White	0.53	0.98	1.0	0.52	0.37	2.26	0.25	18.75	69.17
Draw	0.69	0.97	1.0	0.67	0.73	1.54	0.85	32.35	73.5
Overall	0.55	0.98	1.0	0.53	0.53	1.93	0.45	21.91	73.26

Finally, Table 5 reports also the performance on the process prediction. *Avg* is the average position of the correct prediction in the ranking; *Acc* reports the accuracy, computed by normalizing *Avg* (1 meaning the correct process is the first in the ranking, 0 meaning it is the last). Columns *F*, *A*, *L* report respectively, on average, at what point during a case the following events occur: the correct process becomes the *F*irst in the ranking for the first time (possibly sharing the first position with others); the correct process becomes *A*lone the first in the ranking for the first time; the correct process is not the first in the ranking for the *L*ast time (i.e., there are no more incorrect predictions after *L*). Each process performs best on a different parameter: 'black' on *A*, 'white' on *L*, 'draw' on *Acc*. Interestingly, while the overall accuracy is not outstanding, the correct process appears very early at the top of the ranking, and after less than 3/4 of the process enactment on average the correct process is definitively assessed.

6 Conclusions and Future Work

The main objective of this paper was checking whether, and to what extent, advanced process mining techniques, and specifically the WoMan framework for process mining and management, can support efficient and effective knowledge discovery in complex domains. For this purpose, we focused on chess playing, and cast it as a process. The experimental outcomes showed that, albeit further work is to be carried out, satisfactory initial results have been obtained for all objectives. From the process mining perspective, WoMan proved able to learn chess models effectively and efficiently. From the chess perspective, the learned models discovered interesting features of the game, including (fragments of) some well-known notions and techniques in the chess literature.

As a future work, we will investigate the other features of WoMan, to check whether additional useful information can be discovered (e.g., it may learn preconditions that correspond to the rules of the game or to relevant game strategies). We also plan to find other domains that provide complex processes from which trying to discover useful information.

Acknowledgments. This work was partially funded by the Italian PON 2007–2013 project PON02_00563_3489339 'Puglia@Service'.

References

1. Agrawal, R., Gunopulos, D., Leymann, F.: Mining process models from workflow logs. In: Schek, H.-J., Alonso, G., Saltor, F., Ramos, I. (eds.) EDBT 1998. LNCS, vol. 1377, pp. 467–483. Springer, Heidelberg (1998). doi:10.1007/BFb0101003
2. Cook, J.E., Wolf, A.L.: Discovering models of software processes from event-based data. Technical report CU-CS-819-96, Department of Computer Science, University of Colorado (1996)
3. Cook, J.E., Wolf, A.L.: Event-based detection of concurrency. Technical report CU-CS-860-98, Department of Computer Science, University of Colorado (1998)
4. de Medeiros, A.K.A., van Dongen, B.F., van der Aalst, W.M.P., Weijters, A.J.M.M.: Process mining: extending the α-algorithm to mine short loops. In: WP 113, BETA Working Paper Series. Eindhoven University of Technology (2004)
5. Ferilli, S.: Woman: logic-based workflow learning and management. IEEE Trans. Syst. Man Cybern. Syst. **44**, 744–756 (2014)
6. Ferilli, S., Esposito, F.: A logic framework for incremental learning of process models. Fundam. Inform. **128**, 413–443 (2013)
7. Herbst, J., Karagiannis, D.: An inductive approach to the acquisition and adaptation of workflow models. In: Proceedings of the IJCAI 1999 Workshop on Intelligent Workflow and Process Management: The New Frontier for AI in Business, pp. 52–57 (1999)
8. Lai, M.: Giraffe: using deep reinforcement learning to play chess (2015). CoRR, abs/1509.01549
9. Lloyd, J.W.: Foundations of Logic Programming, 2nd edn. Springer, Heidelberg (1987)
10. Oshri, B., Khandwala, N.: Predicting moves in chess using convolutional neural networks. In: Stanford University Course Project Reports - CS231n: Convolutional Neural Networks for Visual Recognition, Winter 2016
11. Pesic, M., van der Aalst, W.M.P.: A declarative approach for flexible business processes management. In: Eder, J., Dustdar, S. (eds.) BPM 2006. LNCS, vol. 4103, pp. 169–180. Springer, Heidelberg (2006). doi:10.1007/11837862_18
12. van der Aalst, W.M.P.: The application of petri nets to workflow management. J. Circuits Syst. Comput. **8**, 21–66 (1998)
13. van der Aalst, W.M.P., Weijters, T., Maruster, L.: Workflow mining: discovering process models from event logs. IEEE Trans. Knowl. Data Eng. **16**, 1128–1142 (2004)
14. Weijters, A.J.M.M., van der Aalst, W.M.P.: Rediscovering workflow models from event-based data. In: Hoste, V., De Pauw, G. (eds.) Proceedings of the 11th Dutch-Belgian Conference of Machine Learning (Benelearn 2001), pp. 93–100 (2001)
15. Wen, L., Wang, J., Sun, J.: Detecting implicit dependencies between tasks from event logs. In: Zhou, X., Li, J., Shen, H.T., Kitsuregawa, M., Zhang, Y. (eds.) APWeb 2006. LNCS, vol. 3841, pp. 591–603. Springer, Heidelberg (2006). doi:10.1007/11610113_52

Author Index

Printed in the United States
By Bookmasters